AF361457

CYBER WARGAMING

CYBER WARGAMING

RESEARCH AND EDUCATION FOR SECURITY IN A DANGEROUS DIGITAL WORLD

FRANK L. SMITH III, NINA A. KOLLARS, AND BENJAMIN H. SCHECHTER, EDITORS

GEORGETOWN UNIVERSITY PRESS / WASHINGTON, DC

The publisher is not responsible for third-party websites or their content. URL links were active at time of publication.

Library of Congress Cataloging-in-Publication Data

Names: Smith, Frank L., III, 1978– editor. | Kollars, Nina A., editor. | Schechter, Benjamin, editor.
Title: Cyber wargaming : research and education for security in a dangerous digital world / Frank L. Smith III, Nina A. Kollars, and Benjamin H. Schechter, editors.
Other titles: Research and education for security in a dangerous digital world
Identifiers: LCCN 2023010960 (print) | LCCN 2023010961 (ebook) | ISBN 9781647123956 (paperback) | ISBN 9781647123949 (hardcover) | ISBN 9781647123963 (ebook)
Subjects: LCSH: War games. | Cyberspace operations (Military science)
Classification: LCC U310 .C93 2024 (print) | LCC U310 (ebook) | DDC 793.9/2—dc23/eng/20230725
LC record available at https://lccn.loc.gov/2023010960
LC ebook record available at https://lccn.loc.gov/2023010961

24 23   9 8 7 6 5 4 3 2   First printing

Printed in the United States of America

Cover design by Jeremy John Parker
Interior design by BookComp, Inc.

CONTENTS

PART II: EDUCATIONAL GAMES

PART III: CONCLUSION

ILLUSTRATIONS

TABLES

FIGURES

ACKNOWLEDGMENTS

Important contributions to this book extend far beyond the pages that follow. Our original inspiration is owed to the Cyber & Innovation Policy Institute (CIPI) at the US Naval War College. CIPI is a remarkable group of people, past and present. This book was only possible because of support from the CIPI team, including team members who did not write chapters but nevertheless provided invaluable insight and advice along the way, particularly Sam Tangredi, Michael Poznansky, Jason Vogt, Daniel Grobarcik, and Peter Dombrowski.

In addition, we are exceptionally grateful to the contributing authors, without whom there would be no book. Writing a chapter for this volume was a demanding task, given both the content and circumstance. It is hard to write clearly about cyberspace and wargaming. It is harder still to do so during a global pandemic. The contributing authors not only shared their expertise but also their patience. We want to thank each of them, along with Don Jacobs at Georgetown University Press, for investing in this endeavor and seeing it through.

We also want to thank colleagues in the Wargaming Department and across the Naval War College who have helped expand and refine how we think about our shared interests. The same is true for instructors in the Military Operations Research Society's certificate program in cyber wargaming. We are also grateful for generous support from the Naval War College Foundation for education, research, and outreach by CIPI.

Most important, we are indebted to our friends and family for their love and support.

The usual disclaimer applies. The ideas expressed in this book are those of the individual authors: they do not necessarily represent the views of the US Naval War College, the US Navy, the US Department of Defense, the US government, or any other institution.

1

SHALL WE PLAY A GAME? FUNDAMENTALS OF CYBER WARGAMING

FRANK L. SMITH III, NINA A. KOLLARS, AND BENJAMIN H. SCHECHTER

Yes, cyberspace is critical for security. Why, when, and how are not always clear, however. Because of this unsettling combination of significance and mystery, the need for good decision-making about cyber—policy, strategy, operations— has never been more acute. This is true for government, industry, academia, and society at large.

As cyberspace has become central to our social, political, and economic lives, it has become rife with costly and dangerous conflict. For example, Russia interfered in the US presidential election in 2016, combining computer hacking with influence operations to destabilize American democracy.[1] Russian malware named NotPetya also spread from Ukraine around the world in 2017. This attack caused $10 billion in damage to companies such as Maersk, Merck, and FedEx, using an exploit that had leaked, ironically, from the US National Security Agency.[2] The list of major cyber incidents goes on and on, from the SolarWinds breach by Russian intelligence in 2020, to the hack of Microsoft Exchange by China in 2021, to the ransomware attack on Colonial Pipeline by Russian cyber criminals a few months later. No doubt other incidents will have occurred by the time you read this book.

Cyber threats are not new. Nor is conflict in cyberspace. But practical knowledge about them is lagging. Unlike land, sea, air, or outer space, cyberspace is a human creation; people built it, and yet the complexity of our creation seemingly defies comprehension. The scale, scope, and speed of networked information technologies provide fertile ground for unexpected vulnerabilities, interdependencies, and knock-on effects. Surprises lurk online and off, ranging from shocks to global supply chains for hardware and software to the myriad helpful and harmful ways that people use cyberspace. Information technologies make it possible to share information about threats. But that does not mean information sharing actually happens. Authoritative data about cybersecurity are often fragmented and inaccessible (e.g., as private, protected, or secret information).

The net result is that decision-makers in government, industry, and civil society struggle to understand cyberspace in a comprehensive fashion. Researchers, educators, and students in academia are no different. Fundamental questions remain unanswered: How should we conceive of cyberspace? What are viable offensive and defensive actions in this domain? What effects—or returns on investment—do these actions have? And how do cyber effects influence other aspects of business or war? Carl von Clausewitz famously described war as being so interactive, so complex, and so confusing that it inevitably creates "friction" between the best-laid plans and actual practice. Much the same can be said about cyberspace. *Cyber friction and fog make it difficult for policymakers, practitioners, and scholars to answer important questions about what to do.*

Cyber wargames can help. Wargaming is a powerful approach to solving interactive, complex, and confusing problems. "Wargames can save lives," according to Matthew Caffrey, and "make the difference between victory and defeat."[3] For example, wargames were the centerpiece of innovative thinking at the US Naval War College during the 1920s and 1930s. These wargames helped American military planners successfully prepare for key battles against Imperial Japan during World War II.[4]

Even references to wargames in popular culture have had serious consequences. In the 1983 movie *War Games*, a "tech-whiz teenager" hacks a military computer linked to nuclear command and control.[5] The artificial intelligence (AI) on this computer asks the hacker, "shall we play a game?" And their game nearly triggers World War III. Farfetched as this plot may sound, the movie shaped policy in the real world. Watching it inspired President Ronald Reagan, a former movie star, to ask his military advisers if such a nightmare scenario could happen. Finding that it could, Reagan signed a new national security decision directive on telecommunications and automated information system security in 1984 (NSDD 145). The US Congress also started creating new laws on cybersecurity, including what became the Computer Fraud and Abuse Act of 1986.

Cyber wargames are not only a plot line for technothrillers, and wargaming is not only about war—despite its name. Wargames are about human decision-making and the consequences of those decisions, particularly in the face of conflict and uncertainty.[6] This means that wargames are a general-purpose tool. They can be used for both research and education on a wide range of topics, from nuclear deterrence and future warfighting to crisis management and pandemic response.

Contrary to popular belief, cybersecurity is also about human decision-making: not just hardware, software, networks, and data. The combinations of social and technical knowledge required to understand cyberspace—and the countless ways in which it relates to security—do not fit well within traditional academic disciplines, such as political science and computer science. Nor do best practices for cybersecurity fit neatly within professional distinctions, say, between the c-suite and IT department. Fortunately, as a general-purpose tool, wargaming is interdisciplinary. When used correctly, cyber wargaming can bridge the gaps between social and technical knowledge in university class-rooms, corporate boardrooms, and military headquarters.

Cyber wargames can be played in any of these different settings. You do not need a high-tech venue to play out decision-making about advanced tech-nology. While some cyber wargames are played on computers, others resemble classic board games, with dice, cards, and tokens for the players. The number of players can range from two people to dozens or even hundreds of participants. Small games can be played in a single room. Larger games require more space. Or they can be played remotely as a video conference. They can be played over the course of a few hours, as typically is the case in academia and industry, or for several days, as with some cyber wargames in the government and military. As a tool, they are useful in part because they are flexible.

Alas, despite their utility and flexibility, cyber wargames are often shrouded in mystery. This book is among the first to explain how wargames can advance research and education on cyber policy, strategy, and operations.[7] Whereas many cyber wargames are secret or cloistered away inside government and industry, this book openly describes important examples in plain English. Whereas many cyber experts do not interact with wargamers, this book brings together innovative voices from across professional military education, civilian agencies, private industry, think tanks, and academia. Whereas many conven-tional wargames eschew or ignore cyberspace, this book explains different ways to think about this domain, actions in it, their effects, and how these effects cross digital and physical worlds.[8]

The lessons learned through cyber wargaming are applicable beyond these games. They shed light on how people interpret, act upon, and experience cyber-space and other emerging technologies in real life. Cyber wargaming therefore provides valuable insight into how individuals and groups really think and act. This book likewise provides valuable insights for policymakers, practitioners, analysts, teachers, and students who are looking for effective ways to answer hard problems about business and war in cyberspace.[9]

WHY WARGAME CYBER?

Wargames are useful for research, education, and entertainment. Admittedly, entertainment is important, but we focus on "serious" or "professional" war-games that do work. While they may be entertaining as well, the primary pur-pose of these games is to create and disseminate knowledge. Be they analytical or educational, wargames do work by addressing human decision-making, uncertainty, and competition or conflict.

Cyber wargames are a special kind of wargame. They address human decisions about cyberspace. We use the term "cyber" broadly to refer to networked digital information technologies. Despite their technical content, cyber wargames are about people rather than technology assessment. The players are human. They have human agency. They also have complex emotions, identities, and interpersonal relationships to boot.

All these human factors relate to the complex and uncertain problems that wargames aim to address. People are not omniscient. We grapple with risk, uncertainty, and indeterminacy, especially when we are trying to make decisions about complex and dynamic environments such as cyberspace. Here, multifaceted and shifting stacks and layers of technology blend with the behaviors and beliefs of billions of users and organizations. Cyber wargames explicitly grapple with social and technical features of this domain—unlike narrow technical challenges, such as "capture the flag" competitions, which typically ignore social context. By addressing human factors (including group dynamics), wargames can provide traction on ill-defined problems and the possibility of surprise. As Thomas Schelling once observed, "one thing a person cannot do, no matter how rigorous his analysis or heroic his imagination, is to draw up a list of things that would never occur to him."[10] Cyber wargames can help reveal gaps and errors in how we think.

Just as wargames are not only about war, cyber wargames are not only about cyber war. They are about competition and conflict, however. All cyber wargames address some sort of struggle, even when they also include cooperation. In some wargames, players struggle against intentional—perhaps malicious— actions by other people, organizations, or governments. In other games, players struggle against competing demands on limited resources as they respond to accidents, errors, and natural disasters.

These struggles play out in the abstract. Wargames are not field exercises with real weapons or military forces. Nor are they a cyber range for penetration testing or technical training with real equipment and code.[11] Instead, wargames focus on what people might decide to do when faced with adversity and the consequences of their prior decisions. How might decision-makers react when their organization is hacked, for example? Or when their government hacks another country that then hacks them back? Cyber wargames are useful tools for examining these vexing questions.

HOW TO WARGAME CYBER?

There is no single way to wargame. This book answers the question of how to wargame cyber by providing a broad array of instructive examples, coupled with candid reflections on these games from the researchers and educators who created them. Reading this book does not require deep expertise in cybersecurity or wargaming. Instead, we seek to showcase this hybrid field in a way that is accessible to everyone—regardless of whether you are a seasoned expert or someone who is simply curious. The chapters in this book have been organized to clarify the murky and troublesome aspects of cyber wargaming.

With this introduction in hand, you are welcome to read the volume cover to cover or jump to the chapters that you find most interesting. Newcomers to cybersecurity or wargaming should note that this volume builds upon rather than summarizes large bodies of literature in each of these fields.[12] We focus on their intersection. As we see it, cyber wargames vary widely, and yet some of the basic issues are universal. These include game objectives, human players, cyber scenarios, actions, and adjudication.

Game Objectives: Play with Purpose

Cyber wargames, like all wargames, are tools. Whether these tools are useful depends on the task at hand. Everything else about a wargame—its players, scenario, actions, adjudication—depends on the objective or purpose. Form should follow function. If it does not, then the wargame stands to fail. Sadly, many do. In the Department of Defense, for instance, the notion of a "Bunch Of Guys Sitting Around a Table" has earned the acronym "BOGSAT," which is one way that wargames fail when they lack clear objectives.[13]

The most important question to ask about cyber wargaming is "why wargame?" Or, said another way, "what's the point?" This book focuses on two basic objectives—specifically, cyber research and education. Granted, players can learn in research games and educational games can inspire new research. Nevertheless, it is useful to define the purpose of any wargaming in terms of these related but distinct objectives: is the main point to help educate the players, or is it to help researchers study decision-making?

Here the objectives of research are to create new knowledge by discovering, exploring, or analyzing solutions to problems with cyber. For some problems, research objectives can be accomplished through experimental or analytical wargames.[14] Useful examples include iterated wargaming research with the *International Crisis War Game*, discussed in chapter 3; David Banks and Benjamin Jensen's *Island Intercept* and *Netwar* games, discussed in chapter 4; Chris Dougherty and Ed McGrady's remote wargaming, discussed in chapter 5; and the other analytical games discussed in the first half of this book.

Another objective is education—namely, to impart knowledge about cyberspace and security therein. For some lessons, learning objectives can be achieved through educational or experiential wargames.[15] Game-based learning can be engaging. Educational games are widely recognized in studies on pedagogy, not only for being more effective than stand-alone lectures and rote memorization but also for providing a stronger understanding of context in complex environments.[16] Wargames are frequently used as capstone events in professional military education, as described by Benjamin Leitzel in chapters 13, as well as by Hyong Lee and James DeMuth in chapter 14. Cyber wargames are also popular among students at civilian universities, as described by Safa Shahwan and Frank Smith in chapter 9. Educational wargames can teach players about conflicts that they can fight and learn from without the loss of real blood and treasure.

More specific objectives fall under the general headings of research and education. Good wargames are designed accordingly. More often than not, the more precise the objectives are for a game, the better the outcomes. Research games hinge on clearly defined analytical logics that resemble the scientific method. When well designed, the learning objectives for educational wargames are clear as well. These objectives may range from creative thinking and problem solving, as several chapters suggest, to cyber defense and emergency response training, as in Matt Duncan's account of *GridEx* in chapter 10.

Wargames are a general-purpose tool. But they are not an all-purpose tool. A cyber wargame does not test whether criminals can penetrate a law firm's firewall, for example, or whether spies can snoop on government data stored in the cloud. In chapter 2, Andrew Reddie, Ruby Booth, Bethany Goldblum, Kiran Lakkaraju, and Jason Reinhardt explain the trade-offs that researchers must make when choosing between wargames and alternative methods. Likewise, in chapter 8, Andreas Haggman considers what playful learning about cybersecurity both can and cannot accomplish for education. Given the capabilities and limitations of this tool, the most important step in successful cyber wargaming is to specify your objectives and determine whether a wargame is well suited to addressing them in the first place.

Players: Human Factors

Because cyber wargames address human decision-making, the human players are paramount.[17] Questions about who plays should be answered based on the game's objective. For example, if the objective is to improve incident response and resilience in the private sector, then executives from the c-suite are probably the best players, as Maxim Kovalsky, Benjamin Schechter, and Luis Carajal-Kim describe in chapter 11. If the objective is to teach military officers about cyber operations and doctrine, then the most relevant players are probably students at schools such as the Army, Navy, and Air war colleges. Either way, the stated objective for a game may differ from the objectives of individual players, who may have other motives, as well as very human biases and social relationships, as Rachael Shaffer describes in chapter 7.

The relationship between players and cyber wargames may differ from wargames that do not address cyberspace in at least two respects. First, the roles that individuals or teams play in cyber wargames may differ (i.e., who plays what). Like the players themselves, the relevant roles—positions or characters— depend on the objective or purpose of the game. Potential roles range from "chief executive officer" and "chief technology officer" in cyber wargames about private industry to "national security adviser" and "combatant commander" in games about military operations. Decision-making about cyberspace often involves different people and organizations, and thus different roles or positions, from decisions made about other aspects of government or business.

Cyberspace also extends across legal, organizational, and geographic borders, among others. Consequently, the range of players and roles that are relevant to cyber wargames may be more diverse than conventional wargames focused on, say, sea control. Rather than just "blue" and "red" teams facing off against each other, cyber wargames may also warrant "green" and "gray" teams that represent allied or unaligned but nevertheless influential actors in cyberspace. Similarly, cyber wargames for the military may need to include civilian players or roles for private industry and civil society, just as cyber wargames for industry and civilian agencies may need to include military players or roles.

Second, the expertise that players bring to cyber wargames may differ from other games (i.e., what each player already knows). Despite the ubiquitous cyber awareness training that pervades the modern workplace, most people are not cyber experts. Nor are most leaders or policymakers. Some are, however. The wide variation in background knowledge that players bring to cyber wargames can have a significant impact. Are the players in your game undergraduate students, network administrators, corporate lawyers, senior officers, or a mix?

Some players' naïveté about the cyber domain is qualitatively different from their potential ignorance about the physical domains of land, sea, and air. Even if you are unfamiliar with infantry maneuvers, naval engagements, or aerial campaigns, you still enjoy some applicable knowledge—tacit and explicit—by virtue of having lived experience with earth, water, and air since you were born. Yet the physics and structure of cyber "terrain" can seem alien, even for people who use the Internet every day. Neither ignorance nor expertise prevents successful cyber wargaming. In fact, these games provide a unique vehicle through which experts and nonexperts can meaningfully work together. But preparing for what players do and do not know is critical for achieving the research or educational objectives of the game.

Cyber Scenarios: "It Was a Dark and Stormy Night…"

Much of what makes a wargame a game, rather than real life, is the fictional scenario or story about the world that it depicts. Ed McGrady describes wargaming as a kind of collective storytelling.[18] Telling stories helps us make sense of our world, and stories about cyberspace are no exception. Said another way, cyber wargames are dramas. Dramatic elements such as the setting and plot provide context. Context is critical for research and education because human decision-making depends on the context. It is also important for player engagement. Without proper context, players may fight the scenario—namely, protest the game's premise and plot—rather than willfully suspend their disbelief enough to play as intended.[19]

Because some aspects of cyberspace are unfamiliar to some players, scenarios that effectively set the stage for cyber wargames are particularly important. On the one hand, all wargames simplify how they represent the real world. No game captures all the minutiae of sea, land, air, space, or, for that matter, cyberspace. What you include and exclude is a critical design decision. Philip Sabin describes this decision as a choice "between accurately capturing the complex details of a real conflict and keeping the simulation simple enough to be understood and played in a reasonable time."[20] However you choose, your representation will always be an abstraction. Ideally, this representation will provide a shared point of reference for the players. But much of the labor of "filling in the blanks" of the scene is left up to the players' knowledge, experience, and imagination.

On the other hand, representing cyberspace presents special challenges. Experts still debate the nature of this domain—including its size, scope, technologies, and users. Representing it in ways that provide common points of reference and comprehensible visuals is not easy. The dynamics of cyberspace do not help. Not only do the hardware, software, networks, and data change over time (sometimes rapidly), but the users also change, as does their behavior. In contrast, there is little need for a naval wargame to specify that the Pacific Ocean will behave the same way throughout the game. In a cyber wargame, the analogous terrain could radically change. It could even disappear. This is roughly akin to declaring that the entire ocean could freeze, or that ships at sea might suddenly teleport to the other side of the Earth.[21] There is a logic to cyberspace: it is not magic or random. But it is complicated. Thus, setting the stage for a cyber wargame warrants careful attention.

Different wargames address these challenges by providing players with different kinds of background information. For example, the *Cyber 9/12 Strategy Challenge* described in chapter 9 provides a rich description of the scenario, which players receive weeks in advance of the wargame. In contrast, the virtual cyber wargame described by Chris Demchak and David Johnson in chapter 12 is intended to be playable with limited preparation, so it only provides a brief description of the conflict scenario immediately before the game begins.

The kind of information provided also depends on the level of decision-making that the game aims to address. Strategic wargames address national interests, as illustrated by the matrix game discussed in chapter 14. These games typically abstract away a lot of technical detail and treat large swaths of cyberspace as a "black box." They focus instead on high-level inputs and outputs to decision-making by government officials who wield diplomatic, informational, military, and economic instruments of national power.

Operational wargames focus on decision-making inside a particular theater, campaign, or area of operation, such as the conflict with China over Taiwan described in chapter 5. More granular information is warranted at the operational level (e.g., the status of warfighting functions, such as command, control, and logistics; or business functions, such as industrial processes, human resources, and public relations).

Tactics addresses specific steps for attack and defense. For example, offensive tactics are often categorized in terms of the "cyber kill chain," starting with reconnaissance and weaponization, followed by delivery and exploitation.[22] There are tactical cyber wargames.[23] Yet the technical details involved sometimes squeeze out consideration of other human factors and, with them, the basic ingredients of a wargame. This book focuses instead on operational and strategic games. The narratives used to describe scenarios at these levels will vary, as will the maps and other graphic aids provided to help players visualize physical geography and cyber terrain.[24]

Action: How Do You Cyber?

Much of what makes a wargame a "cyber wargame" are the different kinds of actions that are possible in cyberspace.[25] Information is to cyberspace what soil is to land, water is to sea, atmosphere is to air, and vacuum is to outer space. Cyberspace is connected to physical space through the hardware that it runs upon, as well as through the infrastructure, operational technologies, and machinery that it controls (e.g., your car). But the content, volume, flow, and distribution of information obey different laws than physical objects. Force in and through cyberspace differ as well.

As a result, cyber offense and defense are different than attacking and defending physical territory. Offensive and defensive tactics—ranging from spoofing, malware injection, denial of service, and buffer overflow attacks to firewalls, intrusion detection, antivirus, and backups—are special kinds of actions and somewhat unique to cyberspace. Attributing these actions to specific actors can also be difficult, due in part to the wide range of state and nonstate actors involved. "Who acts?" is not a trivial question. Granted, soldiers and spies conducted secret information and influence operations long before the Internet. Nevertheless, these operations can have distinct characteristics and consequences when they are conducted through cyberspace.

One of the most interesting challenges in wargaming is how to best represent the range of cyber actions that are available. Action at a distance is especially interesting because distances differ in cyberspace: both in terms of time and space. Physical distance is far less of a constraint on action through networked information technologies than it is in physical domains. Cyber actions and effects can be local and global. Representing the relevant features of the domain differs from depicting physical geography with the hex maps and charts commonly used in other kinds of wargames.

Temporal distance is another issue. Clichés about "machine speed" and "digital speed" aside, only some cyber actions are fast. Others are slow. For instance, studies of the Stuxnet attack discovered in 2010 suggest that the lead time to develop and deploy a sophisticated cyber exploit may be months or years.[26] Likewise, the term "advanced persistent threat" speaks to the depressingly slow speed at which many intrusions are detected and removed. But persistence is not guaranteed. Some cyber exploits are perishable. They can expire with a software patch or update in ways that bullets, bombs, and missiles do not. For example, the incentive to "use or lose" a malicious exploit seems to have prompted the hackers who targeted Microsoft Exchange to drastically increase the scale of their attack before the vulnerability was rendered obsolete by a software patch in 2021. These temporal dynamics can be gamed, as Paul Schmitt, Catherine Lea, Jeremy Sepinsky, Justin Peachy, and Steve Karappi describe in chapter 6.

The range of possible moves and countermoves is so large that they can become daunting or distracting. This risk is most evident when cyber actions are only intended to be part of a wargame that involves other kinds of diplomatic, military, and economic actions (i.e., wargames with cyber). That said, how much cyber action gets included is a matter of degree. It is not all or nothing, and the vast possibilities are also challenging in wargames that explicitly focus on this domain (i.e., wargames about cyber). Left to their own devices, players risk disappearing down the rabbit hole of their own idiosyncratic machinations about imaginary hacks and defensive options.

Fortunately, there are several ways to constructively constrain the decision space. As with the International Crisis War Game described in chapter 3, analytical wargames that test specific hypotheses can control for confounding variables by using scripted scenarios, in which case players act in response to predetermined updates and injects. Educational wargames can do the same, as with the *Cyber 9/12 Strategy Challenge* described in chapter 9. Alternatively, players can be given a preset menu of cyber actions or, similarly, play with a preset deck of cyber cards (analogous to "spell cards" in entertainment games like *Dungeons & Dragons*). Of course, there are costs to constraining the decision space (e.g., diminished realism), and decisions about these trade-offs in game design affect gameplay as well as adjudication.

Adjudication: Keeping Score of Cyber Effects

When players act in a wargame, the effect of that action is determined by adjudication. Did a player's cyberattack disrupt their opponent's logistics, for example? How much, and for how long? Different games have different methods. They range from "rigid" adjudication using fixed rules to "free" adjudication by subject matter experts or the players themselves. Any way you do it, adjudication depends on a theory of the case, namely, an implicit or explicit model of how and when players' actions cause potential outcomes.

To be clear: wargaming is not the same thing as modeling (although using AI as a player may blur this line).[27] Models and simulations are related but distinct from wargames. Models use logical concepts, math, or mechanical devices like dice to distill the description of a given object, relationship, or phenomenon. Typically, simulations are computer programs that plug data into mathematical or logical models to calculate artificial outcomes. Unlike a wargame, these simulations do not involve human players or the complexities that real people bring to the table. How people act while playing wargames are kinds of data, however. And these data can be fed into implicit or explicit models to decide outcomes in the game.

Wargame models and the data they use for adjudication can be qualitative or quantitative. Numerical data are often privileged. But quantifying cyber effects is difficult, despite the domain's digital content (i.e., 0s and 1s, which are potentially countable) and its human construction (which is potentially understandable). In contrast, quantitative models and metrics for kinetic effects are well established. While not necessarily complete or correct, they are widely accepted as authoritative, which makes them convenient conventions for adjudicating non–cyber wargames with numbers (e.g., the probability of detecting a particular type of missile, or its probability of destroying a particular type of airplane).[28] Cyber effects can be inconvenient. For instance, there is no authoritative cyber version of the *Joint Munitions Effectiveness Manual*, as discussed in chapter 6.[29] Judging how cyber effects interact with other domains further compounds the challenge of adjudication.

These challenges are not insurmountable, however. Nor is quantification the same as scientific rigor (e.g., people make up numbers all the time). Tactical details, such as the performance of a particular missile, are less relevant at the operational and strategic levels of war. In real life, tactical success does not simply aggregate into victory—winning a battle does not mean winning the war. As a result, operational and strategic wargames do not necessarily require granular data about tactics. As with cyber scenarios, you can simplify reality for some aspects of cyber adjudication. Sometimes rolling dice will suffice to decide if a player's hack worked or if their opponent's firewall held.[30]

Subject matter experts can also help adjudicate. Expertise varies, of course, as does expert option. Using experts to judge a game does not get around the challenge of modeling cyber; it just leverages their mental models. As with using dice for probabilistic adjudication, sometimes experts are good enough. Sometimes the players can also weigh in—as with seminar style and matrix wargames, where adjudication is an open discussion or debate. For educational games, adjudication itself can be a learning opportunity, forcing players to think about the efficacy of their actions.

Adjudication is not risk free. It can be controversial.[31] If done incorrectly, it can also cause negative learning, whereby participants and observers learn erroneous lessons from wargaming. Ultimately, the appropriate models and data needed for adjudication depend on the game objectives.

PLAY TO LEARN: CYBER WARGAMING IN RESEARCH AND EDUCATION

Cyber wargames are no panacea. Nor are they a quick fix for research or education. Their objectives, players, scenarios, actions, and adjudication all require careful consideration. Fortunately, many of the questions that arise when designing and playing these games mirror important conceptual and practical challenges in the real world, as do the potential answers.

The subsequent chapters consider these questions and answers in greater detail. The first part of this book focuses on cyber wargaming for research. The second part focuses on education.

Chapter 2 anchors part I of the book on research with a useful framework for understanding trade-offs between contextual realism, player engagement, and analytical utility in game design. Subsequent chapters illustrate alternative approaches to these trade-offs through detailed descriptions of several games. Chapter 3 summarizes the history of cyber wargaming, along with the *Navy Critical Infrastructure War Game* and the *International Crisis War Game*. Chapter 4 describes *Island Intercept* and *Netwar*, and chapter 5 discusses the critical role of imperfect information in operational wargaming. A useful tool for addressing time in cyber operations is examined in chapter 6. The research part concludes with chapter 7, which considers how individual cognition and group dynamics can affect cyber wargame design, gameplay, and research findings as a result.

Turning to educational games in part II, chapter 8 examines the power of playfulness in learning. The next chapters examine different educational games. Chapter 9 describes the *Cyber 9/12 Strategy Challenge*, which teaches students around the world about international crises and cyber policies. Chapter 10 describes *GridEx*, which provides training for critical infrastructure operators and executives across North America. Chapter 11 explains how cyber wargaming can help improve cybersecurity in corporate settings. Since wargaming is prominent in professional military education, the remaining chapters in this part of the book consider cyber wargames for the US military. Chapter 12 is a travel log that describes development of a virtual cyber wargame involving naval conflict. Chapter 14 explains how to teach military doctrine through gameplay, and chapter 15 presents a matrix-style wargame for learning about strategic cyber and information warfare.

We conclude this book by extending the lessons learned for building, playing, analyzing, and learning from cyber wargames to wargaming about other emerging technologies, such as AI and quantum computing. Like software engineers (and hackers), wargame designers borrow good ideas from each other. This book gives you an opportunity to do the same. It also provides citable examples of important cyber wargames and the thinking behind them. Hopefully, this thinking will help you improve your own decision-making about wargaming and cyberspace alike.

NOTES

1. Robert S. Mueller, *The Mueller Report: Report on the Investigation into Russian Interference in the 2016 Presidential Election* (Washington, DC: US Government Publishing Office, 2019), 58; more broadly, see Scott Jasper, *Russian Cyber Operations: Coding the Boundaries of Conflict* (Washington, DC: Georgetown University Press, 2020).
2. Andy Greenberg, *Sandworm: A New Era of Cyberwar and the Hunt for the Kremlin's Most Dangerous Hackers* (New York: Anchor, 2020), 178.
3. Matthew B. Caffrey, *On Wargaming: How Wargames Have Shaped History and How They May Shape the Future* (Newport: Naval War College Press, 2019), 277.
4. John M. Lillard, *Playing War: Wargaming and US Navy Preparations for World War II* (Lincoln: University of Nebraska Press, 2016); Michael K. Doyle, "The US Navy and War Plan Orange, 1933–1940: Making Necessity a Virtue," *Naval War College Review* 33, no. 3 (1980).
5. Fred Kaplan, "'Wargames' and Cybersecurity's Debt to a Hollywood Hack," *New York Times*, February 19, 2016.
6. We define wargames broadly but distinguish them from mathematical models, simulations, and operations research. A narrower but often cited definition for a

wargame is "a warfare model or simulation whose operation does not involve the activities of actual military forces, and whose sequence of events affects and is, in turn, affected by the decisions made by players representing the opposing sides." See Peter P. Perla, *The Art of Wargaming: A Guide for Professionals and Hobbyists* (Annapolis: Naval Institute Press, 1990), 164. Similarly, see Francis J. McHugh, *The United States Naval War College Fundamentals of War Gaming*, 3rd ed. (Newport: Naval War College Press, 2013; orig. pub. 1966).

7. For an earlier compilation of cyber wargames, see John Curry and Nick Drage, *The Handbook of Cyber Wargames* (Bath, UK: History of Wargaming Project, 2020). For some of the first examples of cyber wargaming, see Roger C. Molander, Andrew S. Riddile, and Peter A. Wilson, *Strategic Information Warfare: A New Face of War* (Santa Monica, CA: RAND Corporation, 1996); and Robert H. Anderson and Anthony C. Hearn, *An Exploration of Cyberspace Security R&D Investment Strategies for DARPA: The Day After . . . in Cyberspace II* (Santa Monica, CA: RAND Corporation, 1996).

8. Some military wargames pay lip service to information and cyber operations, but they focus almost exclusively on physical action and kinetic effects. Michael Schwille et al., *Intelligence Support for Operations in the Information Environment: Dividing Roles and Responsibilities between Intelligence and Information Professionals* (Santa Monica, CA: RAND Corporation, 2020), xiv; Christopher Paul, Yuna Huh Wong, and Elizabeth M. Bartels, *Opportunities for Including the Information Environment in US Marine Corps Wargames* (Santa Monica, CA: RAND Corporation, 2020), 12, 35.

9. E.g., see Robert Work, "Wargaming and Innovation," Department of Defense Memorandum, 2015, https://paxsims.files.wordpress.com/2015/04/osd-memowargaming -innovationdepsecdefworkfeb15.pdf.

10. Thomas C. Schelling, "Role of War Games and Exercises," in *Managing Nuclear Operations*, edited by Ashton B. Carter, John D. Steinbruner, and Charles A. Zraket (Washington, DC: Brookings Institution, 1987), 436.

11. As with the definition of wargames, there are different definitions of cyber wargames. For example, the "live fire" NATO Locked Shields Exercise uses simulated computer networks to test operators' skills, which we would categorize as technical training. Still, exercises like Locked Shields are sometimes called cyber wargames. See "Locked Shields," https://ccdcoe.org/exercises/locked-shields/; Curry and Drage, *Handbook*, 21; and Federico Casano and Riccardo Colombo, "Wargaming: The Core of Cyber Training," *Next Generation CERTs* 54 (2019): 88. Others argue that technical wargames are distinct from other technical exercises because of the oppositional and narrative elements. See Bumanglag Kimo et al., "Constructing Large Scale Cyber Wargames," International Conference on Cyber Warfare and Security, 2019, 653.

12. Among others on wargaming, see Perla, *Art of Wargaming*; Christopher A. Weuve et al., *Wargame Pathologies* (Newport: Naval War College Press, 2004); Robert Rubel, "The Epistemology of War Gaming," *Naval War College Review* 59, no. 2 (2006); Philip Sabin, *Simulating War: Studying Conflict through Simulation Games* (London: A&C Black, 2012); Shawn Burns, ed., *War Gamers' Handbook: A Guide for Professional War Gamers* (Newport: Naval War College Press, 2015); James F. Dunnigan, ed., *Zones of Control: Perspectives on Wargaming* (Cambridge, MA: MIT Press, 2016); Elizabeth M. Bartels, "Building Better Games for National

Security Policy Analysis: Towards a Social Scientific Approach" (Ph.D. diss., Pardee RAND Graduate School, 2020); and Sebastian J. Bae, ed., *Forging Wargamers: A Framework for Professional Military Education* (Quantico: Marine Corps University Press, 2022). Likewise, among others on cybersecurity, see Derek S. Reveron, ed., *Cyberspace and National Security* (Washington, DC: Georgetown University Press, 2012); Brandon Valeriano and Ryan C. Maness, *Cyber War Versus Cyber Realities: Cyber Conflict in the International System* (New York: Oxford University Press, 2015); Erik Gartzke and Jon R. Lindsay, "Weaving Tangled Webs: Offense, Defense, and Deception in Cyberspace," *Security Studies* 24, no. 2 (2015); Martin Libicki, *Cyberspace in Peace and War* (Annapolis: Naval Institute Press, 2016); Ben Buchanan, *The Cybersecurity Dilemma: Hacking, Trust, and Fear between Nations* (New York: Oxford University Press, 2016); Adam Segal, *The Hacked World Order: How Nations Fight, Trade, Maneuver, and Manipulate in the Digital Age* (New York: PublicAffairs, 2016); and Tim Maurer, *Cyber Mercenaries: The State, Hackers, and Power* (Cambridge: Cambridge University Press, 2018).

13. Weuve et al., *Wargame Pathologies.*

14. E.g., see Erik Lin-Greenberg, Reid B. C. Pauly, and Jacquelyn G. Schneider, "Wargaming for International Relations research," *European Journal of International Relations* 28, no. 1 (2022): 83–109. We consider experiments, exploration, analysis, and similar activities as variants of research. Whether wargaming itself is a science, an art, or both is a live debate. See Peter P. Perla and E. D. McGrady, "Why Wargaming Works," *Naval War College Review* 64, no. 3 (2011); Andrew W. Reddie et al., "Next-Generation Wargames," *Science* 362, no. 6421 (2018); Reid B. C. Pauly, "Would US Leaders Push the Button? Wargames and the Sources of Nuclear Restraint," *International Security* 43, no. 2 (2018); Ivanka Barzashka, "Wargaming: How to Turn Vogue into Science," *Bulletin of the Atomic Scientists*, March 15, 2019, https://thebulletin.org/2019/03/wargaming-how-to-turn-vogue-into-science/; and Bartels, "Building Better Games."

15. Philip Sabin, "Wargames as an Academic Instrument," in *Zones of Control: Perspectives on Wargaming*, edited by James F. Dunnigan (Cambridge, MA: MIT Press, 2016); Philip Sabin, "Wargaming in Higher Education: Contributions and Challenges," *Arts and Humanities in Higher Education* 14, no. 4 (2015): 329–48; Jeff Appleget, Jeff Kline, and Rob Burks, "Revamping Wargaming Education for the US Department of Defense," Center for International Maritime Security, November 16, 2020, https://cimsec.org/revamping-wargaming-education-for-the-u-s-department-of-defense/; Jean R. S. Blair, Andrew O Hall, and Edward Sobiesk, "Holistic Cyber Education," in *Cyber Security Education: Principles and Policies* (New York: Taylor & Francis, 2020).

16. E.g., see Nina Kollars and Amanda Rosen, "Bootstrapping and Portability in Simulation Design," *International Studies Perspectives* 17, no. 2; Michael D. Kanner, "War and Peace: Simulating Security Decision Making in the Classroom," *PS: Political Science & Politics* 40, no. 4 (2007); Victor Asal, "Playing Games with International Relations," *International Studies Perspectives* 6, no. 3 (2005); and Elizabeth T. Smith and Mark A. Boyer, "Designing in-Class Simulations," *PS: Political Science and Politics* 29, no. 4 (1996).

17. Not all players need to be human. Notional play or artificial intelligence can mimic human behavior, although these options edge toward simulations and modeling rather than wargaming per se. E.g., see A. Walter Dorn, Stewart Webb, and Sylvain

Pâquet, "From Wargaming to Peacegaming: Digital Simulations with Peacekeeper Roles Needed," *International Peacekeeping* 27, no. 2 (2020); and Peter J Schwartz et al., "AI-Enabled Wargaming in the Military Decision-Making Process," Artificial Intelligence and Machine Learning for Multi-Domain Operations Applications Conference II, 2020.

18. Ed McGrady, "Getting the Story Right about Wargaming," *War on the Rocks*, December 30, 2019, https://warontherocks.com/2019/11/getting-the-story-right-about-wargaming/. As it happened, "the word 'scenario' was suggested by Leo Rosten, a comedy screenwriter by day and RAND analyst by night; it came from a genre of Italian theater." Chiara Di Leone, "Imagine Other Futures," *Noema*, February 22, 2022, www.noemamag.com/imagine-other-futures/.

19. Engagement is sometimes called "flow" or "immersion." Having players accept the scenario enough to engage is important, regardless of whether the game is physical or virtual. See David T. Long and Christopher M. Mulch, "Interactive Wargaming Cyberwar: 2025," Naval Postgraduate School, Monterey, CA, 2017.

20. Sabin, *Simulating War*, xxi.

21. Nina Kollars and Benjamin Schechter, "Pathologies of Obfuscation: Nobody Understands Cyber Operations or Wargaming," February 1, 2021, www.atlanticcouncil .org/in-depth-research-reports/issue-brief/pathologies-of-obfuscation-nobody -understands-cyber-operations-or-wargaming/.

22. See www.lockheedmartin.com/en-us/capabilities/cyber/cyber-kill-chain.html.

23. Among others, see Daniel T. Sullivan et al., "Best Practices for Designing and Conducting Cyber-Physical-System War Games," *Journal of Information Warfare* 17, no. 3 (2018): 92–105; Deborah J. Bodeau, Catherine D. McCollum, and David Fox, "Cyber Threat Modeling: Survey, Assessment, and Representative Framework," Homeland Security Systems Engineering and Development Institute, April 2018; Tucker Bailey, James Kaplan, and Allen Weinberg, "Playing War Games to Prepare for a Cyberattack," *McKinsey Quarterly*, 2012, 1–6.

24. Repeated use of similar representations can reify how people interpret space and conflict. See Dan Öberg, "Exercising War: How Tactical and Operational Modelling Shape and Reify Military Practice," *Security Dialogue* 51, nos. 2–3 (2020); also see Walker D. Mills, "The Map Is Never Neutral," *Strategy Bridge*, August 31, 2020, https://thestrategybridge.org/the-bridge/2020/8/31/the-map-is-never-neutral.

25. Categorizing the many different types of cyber attack is a dynamic field. Notably, see MITRE and its Att&ck Matrix: Blake E. Strom et al., "MITRE Att&Ck: Design and Philosophy," technical report, 2018.

26. Rebecca Slayton, "What Is the Cyber Offense-Defense Balance?" *International Security* 41, no. 3 (2016); Jon R. Lindsay, "Stuxnet and the Limits of Cyber Warfare," *Security Studies* 22, no. 3 (2013).

27. Although related, wargaming is not operational research or survey research. Aggie Hirst, "States of Play: Evaluating the Renaissance in US Military Wargaming," *Critical Military Studies*, 2020; Sarah Kreps and Debak Das, "Warring from the Virtual to the Real: Assessing the Public's Threshold for War over Cyber Security," *Research & Politics* 4, no. 2 (2017).

28. E.g., see Lynn Eden, *Whole World on Fire: Organizations, Knowledge, and Nuclear Weapons Devastation*, Cornell Studies in Security Affairs (Ithaca, NY: Cornell University Press, 2004).

29. Also see Mark A. Gallagher and Michael C. Horta, "Cyber Joint Munitions Effectiveness Manual (JMEM)," *American Intelligence Journal* 31, no. 1 (2013); Matthew W. Zurasky, "Methodology to Perform Cyber Lethality Assessment" (Ph.D. diss., Old Dominion University, 2017); and George Cybenko, Gabriel Stocco, and Patrick Sweeney, "Quantifying Covertness in Deceptive Cyber Operations," in *Cyber Deception*, edited by S. Jajodia, V. Subrahmanian, V. Swarup, and C. Wang (Berlin: Springer, 2016).
30. Along similar lines, see Rex Brynen, "Virtual Paradox: How Digital War Has Reinvigorated Analogue Wargaming," *Digital War* 1 (2020): 138–43.
31. E.g., see Julian Borger, "War Game Was Fixed to Ensure American Victory, Claims General," *Guardian*, August 20, 2002.

PART I

RESEARCH GAMES

2

CYBER WARGAMES AS SYNTHETIC DATA

ANDREW W. REDDIE, RUBY E. BOOTH, BETHANY L. GOLDBLUM,
KIRAN LAKKARAJU, AND JASON C. REINHARDT

Imagine a conference room in the Pentagon. In it, a senior policymaker is participating in a cyber wargame. As the "Orange" player, she must decide whether to launch cyberattacks against the military bases of her adversaries. She weighs the actions taken by her fellow players over the past three rounds, including her broken alliance with the "Green" player and the invasion of her territory by the "Purple" player. Will her cyberattacks demonstrate resolve or risk disproportionate retribution? This kind of decision is of serious concern to policymakers, scholars, and game designers alike: namely, under what conditions do actions taken in cyberspace have the potential for conflict escalation?

A significant challenge for researchers who study cyber escalation is the lack of data to test different theories about this critical intersection of politics, security, and technology. The paucity of data fuels long-standing debates over the nature of cyberspace and how it compares with other domains of conflict and competition. Some analysts argue that action in cyberspace is likely to exacerbate the risk of escalating conflict.[1] Others are more sanguine.[2] Unfortunately for researchers—but fortunately for everyone else—empirical data about cyber actions and effects during war remain limited.[3] Reliable evidence from conflicts such as those in Syria and Ukraine are, at best, incomplete. In an attempt to resolve this challenge, researchers have undertaken a variety of efforts to both explore and analyze conflict in the cyber domain by generating synthetic data (i.e., data that is artificially generated, not collected from real-world events). Wargaming represents an increasingly important tool for this endeavor.[4]

In this chapter, we first outline "the wargamer's trilemma" to articulate the trade-offs that wargame designers face. Second, we use this trilemma as a lens to examine the use of wargaming by scholars to study cyber conflict and statecraft. Here we pay particular attention to the wide variety of analytical wargames and scenarios that already exist in the field. Third, we compare wargames with alternative approaches to generate data—specifically, modeling and simulation on one hand and survey experiments on the other hand. In the process, we identify the relative strengths and weaknesses of these different methods—noting where they might usefully augment or complement one another. Based on this comparison, we conclude by highlighting the promise of cyber wargaming to explore and analyze important challenges in national and international security.

THE WARGAME DESIGNER'S TRILEMMA

When creating wargames for the purpose of research or analysis, game designers face a trilemma (i.e., a three-way dilemma).[5] Analytical utility, contextual realism, and player engagement are all in tension. The trilemma concept, often found in economics, refers to complex problems with three mutually exclusive solutions or objectives.[6] We employ this term in a broad sense to describe a decision space in which game designers must make trade-offs between important but at least somewhat incompatible goals.

Our trilemma model illustrates the compromises made when generating synthetic data with wargames, given that there are no perfect solutions.[7] We describe these trade-offs as a game's design moves closer to or further from each node, as outlined in figure 2.1. We acknowledge that our model is a simplification of a complex trade space. However, for the sake of simplicity, we argue that each node and axis is important as well as useful to consider. In the next section, we examine each trade-off in further detail.

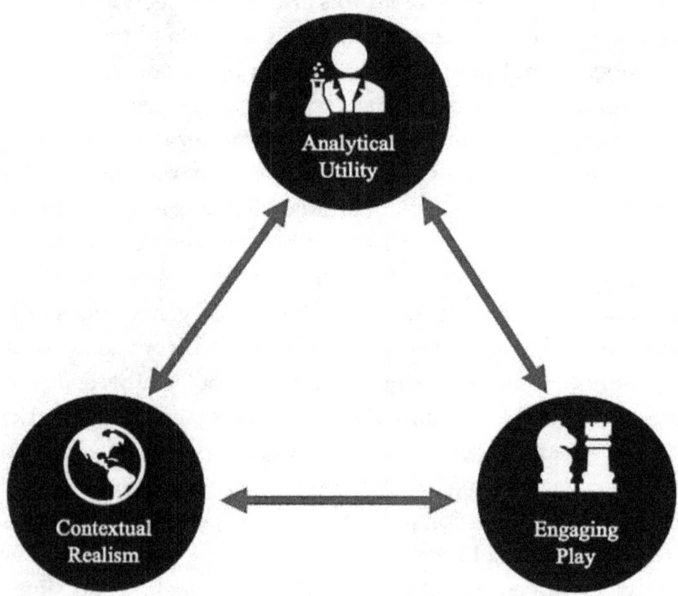

FIGURE 2.1. The Wargame Designer's Trilemma

Balancing Analytical Utility and Contextual Realism

Wargames, like survey experiments and computer-based models, simplify reality to analyze aspects of it. Yet contextual realism is also essential to a wargame's validity as an analytical tool. Contextual realism refers to the extent that a game adheres faithfully to the real world. On one hand, wargame designers must retain some fidelity to "the real world" so that players' behavior during gameplay reflects real-world decision-making. On the other hand, wargame designers also need to simplify reality; otherwise, it is difficult if not impossible to isolate and test how different causes and effects relate to each other. As a consequence, researchers and wargame designers must omit unnecessary or unwanted elements to remove extraneous artifacts that might confound or confuse answers to their analytical question.

This is an important trade-off. Increasing fidelity to the real world, with all its complexity and richness, might confound the analysis. Conversely, by decreasing fidelity through simplification in an attempt to generate and capture only the most relevant data, the designer risks losing essential elements that might drive peoples' behavior in the real world. We are therefore left with the directive: make the game as simple as possible, but no simpler. The difficulty of doing so is one reason why wargaming is sometimes described as an art rather than a science.[8]

Variants of the balancing act between realism and utility are described throughout this book. To strike an appropriate balance, researchers must first identify an appropriate level of fidelity for their analytical objectives; they must then design game mechanics that elicit behavior associated with their research question and measure data relevant to the study variables. This is not an easy task. Simplified design may generate clear measurement but lack important contextual details that, when present, alter behavior. In that case, the game may not generate reliable and valid data related to the phenomenon of interest.

For instance, if a wargame aims to analyze resource management in cyberspace, as is the case in *CyberWar 2025*, and yet lacks a map, it may be possible to measure actions easily; but players may struggle to grasp the significance of their actions without clear visual cues.[9] Too much fidelity to real-world contexts may cause difficulties of a different type. For example, one of our wargames, *SIGNAL*, sought to study escalation in a nuclear context.[10] Working on *SIGNAL*, we found that board design can meaningfully alter a player's behavior, as people apply heuristics from the real world or even fictional contexts to determine appropriate strategies (e.g., the large nation at the top of the map is played as "Russia" and the region to its west as "Europe," despite the absence of such labels). Thus, a balance is required. This need for balance between contextual realism and analytical utility is one of the main tensions in effective wargame design.

Balancing Contextual Realism and Engaging Play

Adding to the challenge, analytical wargame designers must also consider player engagement, wherein the player is "immersed" in a feeling of involvement and energized focus in gameplay.[11] As players become immersed, they are able to "suture" themselves into the game world, merging action and awareness within a state of "flow," which increases their concentration on the task at hand.[12] Engaging play is important for analytical and educational wargames alike. Without engagement, players may be desultory or disinterested. They may make random decisions without weighing the ramifications of their actions, generating unrealistic data. In the real world, we must live with the real consequences of our actions; but in wargames, the consequences are relevant in proportion to a player's investment in the game's outcomes.

As with each node pair of the trilemma, to some extent contextual realism and engaging play are opposed. After all, there are countless minutiae associated with every possible scenario, the boring details of everyday life that do not necessarily add anything to the gaming experience. In a game, however, some of these details may be necessary to construct a sufficiently realistic environment. Engaging games abstract away those facets of life that do not matter to the context while retaining the richness and narrative that give actions meaning.

Imagine that, in addition to implementing high-level policy, our Orange player is forced to choose the rations, uniforms, and bivouac layout for thousands of her troops. Such strict adherence to the concrete details of war would not serve the level of action and reaction—nation versus nation. In fact, that level of commitment to "realism" would render the game unplayable for most audiences—whether driven by a lack of knowledge associated with such fine details or pedantry leading to a lack of engagement. By contrast, one can easily imagine a tactical, squad-based game using a few of those same elements to help the player feel connected to "their" troops. Reaching an appropriate balance between realism and engaging play requires building the narrative of the game at a level of abstraction appropriate for the research question.

Here again, some wargamers argue that designing for engaging play is an art that cannot be codified. While there may be some aspects of game design that are truly ineffable, we suggest that it is possible to identify characteristics that tend to lead to player engagement. Games with a compelling "skin," or narrative, are generally more enticing than those in which the skeleton of their mechanics is visible. In wargaming, an attractive board, a well-drawn map, small narrative touches, and a coherent theme all contribute to immersion, or suture. By contrast, an overly simplistic or amateur design can prevent players from ever engaging or pull them out of the game, exacerbating careless or unconsidered play due to limited investment in game outcomes. This is particularly problematic given that one advantage of wargames as an analytical tool compared with alternative data generation processes—discussed below—is their potential to elicit "authentic" behavior from participants. Wargames rely on emotional investment in game outcomes to elicit realistic decision-making during play.[13] Achieving a certain threshold of investment—whether via win conditions, consequential play, or strategic interaction, along with the game venue and identity of players—is necessary for play that generates useful insights into the questions at hand.

Balancing Engaging Play and Analytical Utility

The final trade-off is between engaging play and analytical utility. From the perspective of analytical utility, the best game is often the most parsimonious (e.g., a simple game with as few decision points as possible that only capture the behavior of interest). However, games that are too simple will not be sufficiently engaging. This tension stands to be particularly acute in educational cyber wargames, where too much analytical utility can detract from player engagement and thus learning.

All things being equal, the more options for action that are available in a wargame, the larger the sample required to achieve statistically significant results. Adding to this difficulty, more prompts, scenarios, and even cosmetic alterations may require more players and more playthroughs for researchers to assess the robustness and validity of the game as a scientific tool. Too broad an option space can confound analytical utility. Too narrow a space can result in a game that no one cares enough about to play. While there are methods to increase player participation through "crowdsourcing" (through sites such as Amazon Mechanical Turk or university laboratories), engagement is still necessary to give validity to the data generated by their play.

The complexity of game mechanics, such as cards and dice, can also affect engaging play. As is true of a player's emotional investment, mechanical complexity has an optimal zone. However, there is no ideal amount of complexity or parsimony. The appropriate level of parsimony depends on the game objective, as well as the players and their knowledge and experience. Hobbyists who have played many games for entertainment may find more complex rules more engaging than policymakers, for example, who may recoil. Thus, it is important to know who will play the game, why, and—of course—to playtest or beta test the game in advance.

As is true with wargaming in general, so too with cyber wargaming. The trade-offs highlighted by this trilemma are no less significant when wargames are used to generate new data and knowledge about cyberspace. In the next section, we examine how a few cyber wargames address the wargamer's trilemma.

WARGAMING THE CYBER DOMAIN

Let us return to our Orange player in the Pentagon. She finds herself in her strategic conundrum based on the scenario that she and her fellow players have been provided—made real in part by the map and limited resources at their disposal. This begs the question of how a different map or different distribution of capabilities might affect player behavior. The "laboratory effects" driven by design considerations could be significant.[14] For example, are the outcomes of games in which players pick from prescribed options—or menu—for cyber actions analogous to games that allow an open response? Similarly, what elements are salient when one seeks to map and model a cyber environment?

To engage with these questions, we use the trilemma outlined above to briefly consider three analytical cyber wargames on escalation dynamics. We consider how the respective games create the conditions for meaningful play—paying special attention to the cyber aspects of the game.[15]

The International Crisis War Game

As described in chapter 3, the *International Crisis War Game* generates synthetic data to examine how cyberattacks on nuclear command, control, and communication may affect the probability of nuclear war.[16] In this analytical wargame, the researchers assign players' specific roles in a national security cabinet. Working together in these roles as a team, the players respond to a crisis scenario involving an adversary and a third party. The researchers manipulate specific aspects of the scenario as an experimental treatment and then measure how the players' teams respond with regard to nuclear weapons. The game is played over three hours by elite players who possess nuclear and cyber expertise.

The *International Crisis War Game* is intended to answer a specific research question and is expressly designed with analytical utility in mind.[17] As per our trilemma, these choices have ramifications for contextual realism and engaging play. In terms of contextual realism, the game does not involve real-world countries or geographies. This choice is understandable given the proclivity of players to caricature scenarios when faced with environments that approximate the real world. Despite the fictional countries and map, however, players are provided with detailed information about the countries involved. This information attempts to provide enough realism to support valid analysis.

The game is also one-sided. Rather than allow for strategic interaction, in which players engage with a live opponent, adversary behavior is prescribed or predetermined. Consistent and repeated adversary behavior simplifies data creation and collection, which serves analytical utility. However, the consequences of players' actions are muted or lost in such one-sided play, departing from the reality in which crises are an interactive—and potentially violent—contest of human capabilities and will. This trade-off risk stands to limit players' engagement. The plot or narrative arc of the game is static. As such, player decisions cannot change the course of gameplay in any meaningful way. Their action in round 1 of the game does not change the scenario in round 2. Therefore, if we are to label the *International Crisis War Game* associated with the trilemma discussed above, it maximizes analytical utility.

Island Intercept

In partial contrast, consider the cyber wargame called *Island Intercept*, which is detailed in chapter 4. This game explores how cyber capabilities affect the likelihood of state aggression.[18] It centers on a South China Sea scenario, beginning with a vignette about Taiwan. Players represent the United States and China. Their actions are adjudicated by an umpire, who uses dice roles to move the game forward.

Unlike the *International Crisis War Game*, with its prescribed scenarios, the players decide who "wins" *Island Intercept*. Each side has cards that describe their specific cyber capabilities (e.g., phishing to gain access to adversary network), as well as their associated costs and benefits (e.g., "political permission level" and "likelihood-of-success measure"). Players decide how to use these capabilities against each other as they work toward the win conditions.

Island Intercept is built with analytical questions in mind: when played repeatedly, it elicits responses for study in a subsequent survey experiment.[19] Yet this cyber wargame hews closely to contextual realism by using real states involved in an all-too-realistic conflict scenario. It also allows for interaction between players, so they must deal with the strategic consequences of their respective actions from round to round—an important aspect of engaging play. Moreover, players select from a wide but not infinite range of possible cyber actions, reducing the complexity of cyber conflict and making the wargame more approachable if not engaging.

These choices about research design might diminish the analytical utility of the game, however. For example, players may caricature the behavior of Washington and Beijing based on their prior knowledge and assumptions, rather than engage with the strategic dilemmas posed by the game itself. In addition, the characteristics associated with the cyber capability cards may influence players' behavior in ways that might not be as externally valid as hoped. Combining this wargame with a survey experiment goes a long way toward addressing these trade-offs. Nevertheless, *Island Intercept* tends to maximize engaged play.

Cyber Escalation Game

Finally, consider the seminar-style wargame that Benjamin Jensen and Brandon Valeriano designed to examine cyber escalation.[20] Similar to the *International Crisis War Game*, but unlike *Island Intercept*, players in this game represent a fictional state in conflict with a nonplayer, peer adversary in a one-scenario, one-sided game. The players receive one of four different briefing packets that represent the experimental treatment. In these packets, Jensen and Valeriano vary whether the player faces a cyber triggering event and whether they have a cyber response option. Data collection seeks to measure whether each player responds in an escalatory manner. This game has been played by over two hundred and fifty participants that included "graduate and undergraduate students, government officials, military officers, and private sector employees."[21]

Jensen and Valeriano use briefing packets to approximate the documents used in real life by the US National Security Council. This choice privileges contextual realism, trading off against engaged play. It is not entirely realistic, however, given the lack of strategic interaction between the players. As a result, this cyber escalation game also tends more toward analytical utility than *Island Intercept*.

In each of these cyber wargames, we see the wargamer's trilemma at work. Each game sits at a different point in the trilemma simplex, as illustrated by figure 2.2. This variation has implications for the data generated by each game.

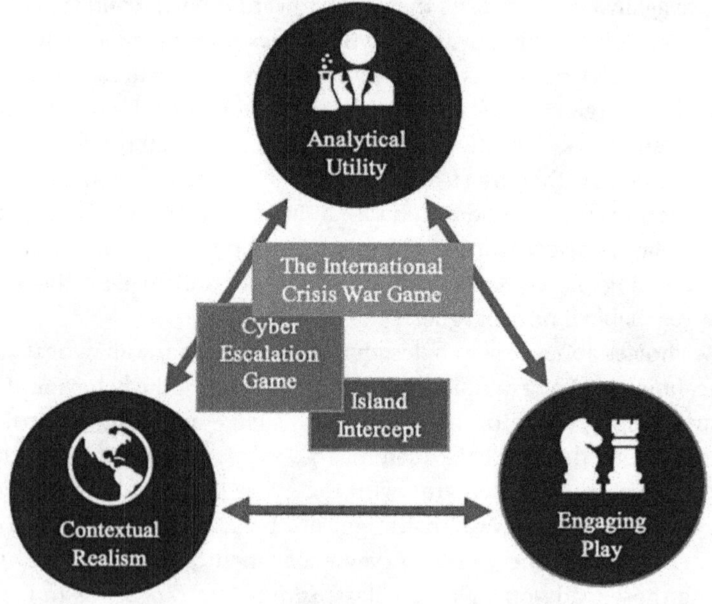

FIGURE 2.2. Positioning Exemplars on the Trilemma's Plane

This diversity highlights the need to further interrogate how data generation through cyber wargaming relates to other research methods.

ANALYZING THE CYBER DOMAIN ACROSS RESEARCH METHODS

Cyber wargames offer an exciting tool for observational and, increasingly, experimental work.[22] So, how does our understanding of the Orange player's decision-making in a wargame compare with other research methods that analysts use to study cyber competition and conflict? Our trilemma provides a helpful framework for comparing and contrasting observational data, modeling and simulation, survey experiments, and wargames. It also helps us identify potential strengths and weaknesses associated with each approach to data generation.

Empirical Data

Scholars typically look to empirical data—whether case-based or statistical—to answer research questions. Questions about cyberspace are no different, although this domain does present empirical challenges. Much of the theoretical work on cyber operations leans heavily on a limited number of cases to argue that the effects of cyber conflict are unique—or not.[23] Extrapolating from these cases, scholars make myriad claims, including suggestions such as that cyberattacks represent a new weapon of mass destruction, that they transform geopolitics due to their relatively low barriers for entry, and that they may offer unique pathways and new motives for conflict escalation.[24]

From our perspective, the challenge with these observational data is twofold. First, wars that involve the overt use of cyber capabilities have been rare, and "cyber war"—where cyberattacks have physical effects—even more so. Said another way, the people involved in a real war are assumed to be fully engaged, due to the existential consequences, but they rarely play their cyber cards in the open. Second, by its nature, observational data are not created with analysis in mind. Indeed, empirical data represent the most extreme example of the tradeoff between contextual realism and analytical utility, as they contain all the noise and extraneous detail of real life. When researchers convert real events into a case study or data set, they must make choices, and these choices reflect not only their own biases but also external biases related to how data are recorded, stored, and presented for public consumption.

As a result, there are important selection effects associated with empirical data. Existing efforts to collect instances of cyberattacks reflect information that is publicly available, which may not provide a full picture given the possibility of covert and clandestine operations. For example, cyberattacks by nonstate actors that use readily available tools may be overrepresented. These attacks and exploits are less likely to be classified, whereas sophisticated state-sponsored attacks are more likely to be kept secret. These selection effects are a concern for analysts seeking generalizable insights. Viewed from the perspective of our trilemma, observational data maximize contextual realism by reflecting events that have occurred, as well as engaging "play," given the high stakes of real life. But they often suffer in terms of analytical utility.

Given these challenges, scholars of cyber conflict have turned to synthetic data-generating processes, which may provide more useful data for isolating the impact of various stimuli. Three popular methods for generating synthetic data include modeling and simulation, survey experiments, and wargaming.

Modeling and Simulation

When applied to cyber, models and simulations range from focusing on a single defender and attacker to multiple agents engaging in cyber war.[25] Modeling and simulation tools can be roughly divided into two related but distinct camps—mathematical models of attacker/defender behavior using game theory, or more complex and adaptive models and simulations of attacker/defender interactions (e.g., agent-based modeling).[26]

Paring strategic problems down to a relatively parsimonious model represents one major strength of these approaches.[27] Put another way, these methods tend to maximize analytical utility. One challenge, however, is that heroic assumptions are often baked into these models (e.g., rationality or perfect information). These assumptions may or may not be externally valid. With such assumptions in place, the models may generate perfect, but perfectly inhuman, results. Simulations attempt to address this shortcoming by adding variables that can be sampled to provide probability distributions; but they can still miss important contextual factors.

There is also a secondary concern regarding the source material used to create the mathematical models in the first place.[28] For example, several modeling approaches rely on inputs from a limited pool of subject matter experts. The resulting models can reflect the proclivities and biases of these experts rather than general truths—once again limiting contextual realism and affecting the analytical utility of the results. This can partly be mitigated by learning from existing data sets, although selection effects still represent an important concern.[29]

Usually, there is no "player" engagement in these models and simulations of strategic interaction since no humans are involved in generating the data beyond the model or simulation designer. Given the types of strategic behavior that scholars and analysts of cyberspace are nominally interested in, models that fail to include human behavior are, in our view, problematic. This has led to efforts to bring humans back into the data-generating process.

Survey Experiments

Survey experiments represent a popular method for including human factors when examining the challenges posed by the cyber domain (and social science in general). They provide only enough contextual realism to observe the relevant behavior. Surveys typically pay little attention to how engaging the scenarios and questionnaires are for respondents.

In survey experiments, researchers randomly assign participants to a treatment or control group. While experimental treatments vary, they are often expressed by either manipulating the scenario—vignette—or by manipulating the response options provided to the participants.

For example, in a vignette-based study, we might vary the military capabilities that players have among their response options. For instance, we might provide some players with offensive cyber capabilities and others with only conventional weapons. Here, we would be concerned with measuring the different choices selected by respondents that have cyber capabilities against those that do not. Unlike modeling and simulation, survey experiments channel the questions posed by the vignette through human decision-making instead of relying upon a mathematical formula or computer program to adjudicate the outcome. The ability to engage with human responses at scale represents a major strength of this method.

A number of scholars have used survey experiments to examine cyber conflict. Some have found that people support the use of cyber weapons, all else being equal.[30] Others have found that their respondents are reticent regarding the use of cyber capabilities.[31] Still others have found contingent support for cyberattacks among the general public—arguing that it depends on attribution, magnitude, or risk perception.[32]

When viewed in terms of the trilemma, survey designers must address three major methodological concerns. The first concern relates to sampling—an integral aspect of analytical utility. Some scholars suggest that making inferences about strategic decision-making requires engaging with policy elites.[33] Others are more sanguine about potential differences between elites and nonelites.[34] While we do not adjudicate this argument here, further research is likely warranted, focused on the sampling effects associated with cyber conflict studies.

The second concern relates to engagement. Survey experiments tend to be short. This limited duration tends to limit the amount of information that surveys can communicate to respondents about the phenomenon under study. Moreover, respondents face minimal consequences for their responses. This limits the engagement or immersion of the survey "experience" and may allow respondents to answer a researcher's questions without weighing the costs and benefits of their actions. Further, evidence suggests that participants tend to respond to survey questions in a manner that reflects their ideals rather than their true behaviors.[35] This bias can be minimized with careful survey design but cannot reliably be eliminated.

Third, survey experiments struggle to incorporate strategic interaction. Many research questions of interest in the cyber domain involve dynamics between at least two actors—if not more. The absence of these interactions in a survey setting limits this method's contextual realism. This limitation is not confined to surveys, of course. As we note above, some cyber wargames also lack strategic interaction. However, for surveys, this limitation is innate to the medium rather than the result of a design choice.

Analytical Wargaming

Wargaming as an analytical method has the potential to balance some of the otherwise stark trade-offs made by surveys as well as modeling and simulation. First, wargames can offer a reasonable approximation of "the real world." As outlined above, cyber wargames can be calibrated to be as complex or as simple as necessary to engage with their respective research questions. Second, wargames incorporate human actions and responses—decisions—that modeling and simulation cannot. They also bring in strategic play between humans that surveys omit. Relatedly, wargames work to overcome concerns regarding the lack of consequences facing participants when they take part in a survey experiment. Players must deal with the ramifications of their actions. As researchers, we may not be concerned with who "wins" or "loses." But the competitive aspect of wargames encourages participant buy-in and, as a result, yields behavior more reflective of real-life situations.

In the simplex bounded by analytical utility, contextual realism, and engaged play, we submit that analytical wargames might represent something of a Goldilocks method. Moreover, they may be used to help overcome some of the weaknesses of any given data-generating process as part of a multimethod research design. Figure 2.3 shows each method discussed above placed on the trilemma's plane. As a whole, analytical wargaming occupies a place in the center. Individual wargames make different trade-offs, as we argue, shifting their placement in one direction or the other. These trade-offs also indicate how best to pair individual wargames with other methods. For example, games designed to hew closely to analytical utility could benefit from pairing with empirical data, while games more focused on engaging play or contextual realism might benefit from a supporting simulation or survey.

THE PROMISE OF CYBER WARGAMING

In this chapter, we outline the wargamer's trilemma and the trade-offs between analytical utility, contextual realism, and engaged play to interrogate both cyber

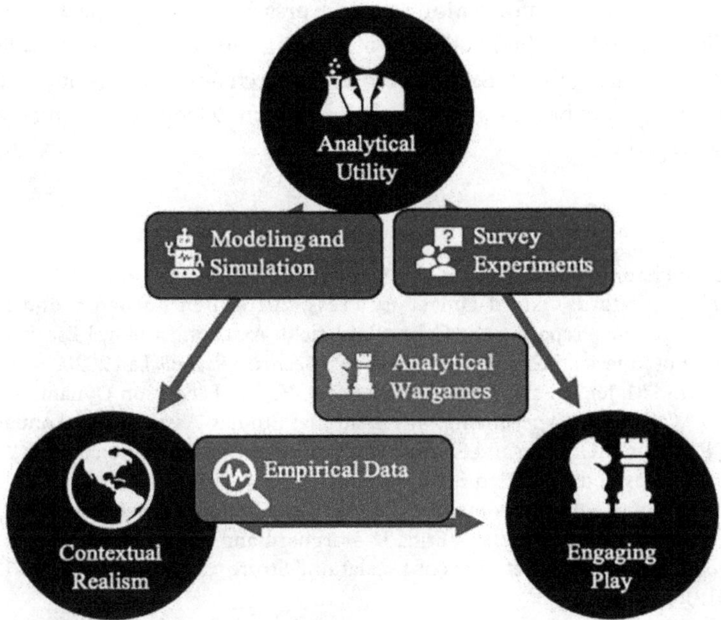

FIGURE 2.3. Methods and the Trilemma

wargames and other methods for generating synthetic data. Our framework reflects the potential of wargaming for cyber research, as well as the need to think critically about the design trade-offs when using wargames as an analytical tool. In our estimation, methodological research that further interrogates the trade-offs described by the wargamer's trilemma has an important role to play in advancing this field's ability to address important research questions beyond the reach of other methods.

In our own work, such as with *SIGNAL*, we use both wargaming and survey methods to examine deterrence in the cyber domain.[36] Specifically, we examine when and under what conditions players make deterrent threats matched with a particular domain, when these deterrent threats succeed, and whether players attempt to deter cyberattacks differently than attacks emanating from other domains. In the process, we have had to weigh the balance between analytical utility, contextual realism, and engaged play—ultimately creating a game that appropriately models both the threat space and the action space.

By extension, we also engage with a renewed scholarly debate concerning the relative costs and benefits of revealing or concealing clandestine—in this case cyber—capabilities to an adversary.[37] This literature is largely theoretical (not surprisingly), reflecting the paucity of open source data on clandestine capabilities in general. Our preliminary results suggest that players treat cyber capabilities differently than conventional and nuclear capabilities—and that the potential to reveal information to an adversary regarding a particular cyber capability may reduce the likelihood of making a threat in the first place. As more scholars engage with this method of research and analysis, it is our hope that the library of cyber wargames and the new knowledge they can create will continue to grow.

NOTES

1. Ben Buchanan, *The Cybersecurity Dilemma: Hacking, Trust, and Fear between Nations* (Oxford: Oxford University Press, 2016); Ben Buchanan and Fiona S. Cunningham, "Preparing the Cyber Battlefield: Assessing a Novel Escalation Risk in a Sino-American Crisis," *Texas National Security Review*, Fall 2020.
2. Benjamin M. Jensen and Brandon Valeriano, "Cyber Escalation Dynamics: Results from War Game Experiments," International Studies Association, Annual Meeting Panel: War Gaming and Simulations in International Conflict, March 27, 2019; Sarah Kreps and Jacquelyn Schneider, "Escalation Firebreaks in the Cyber, Conventional, and Nuclear Domains: Moving Beyond Effects-Based Logics," *Journal of Cybersecurity* 5, no. 1 (2019); Erica D. Borghard and Shawn W. Lonergan, "Cyber Operations as Imperfect Tools of Escalation," *Strategic Studies Quarterly* 13, no. 3 (2019): 122–45.
3. There have been several efforts to capture cyber incident data, such as the Dyadic Cyber Incident and Dispute Data; see Brandon Valeriano and Ryan C. Maness, "The Dynamics of Cyber Conflict between Rival Antagonists, 2001–11," *Journal of Peace Research* 51, no. 3 (2014): 347–60; Center for Strategic and International Studies, "Significant Cyber Incidents," no date, www.csis.org/programs/strategic-technologies-program/significant-cyber-incidents; National Security Archive, "The Cyber Vault Project," February 10, 2016, https://nsarchive.gwu.edu/project/cyber-vault-project; and Council on Foreign Relations, "Tracking State-Sponsored Cyberattacks Around the World," no date. These efforts, though subject to selection effects, are essential to understanding state behavior in the cyber domain.
4. Andrew W. Reddie et al., "Next-Generation Wargames," *Science* 362, no. 6421 (2018): 1362–64.
5. As our colleagues at the Naval Postgraduate School note, wargames are also used for education and training purposes. We focus here primarily upon the analytical applications of wargaming methods: Jeff Appleget and Robert Burks, *The Craft of Wargaming: A Detailed Planning Guide for Defense Planners and Analysts* (Newport: Naval Institute Press, 2020).
6. Dirk Schoenmaker, "The Financial Trilemma," *Economics Letters* 111, no. 1 (2011): 57–59.
7. Joseph E. McGrath, "Dilemmatics: The Study of Research Choices and Dilemmas," *American Behavioral Scientist* 25, no. 2 (1981): 179–210.
8. Peter P. Perla, *The Art of Wargaming: A Guide for Professionals and Hobbyists* (Annapolis: Naval Institute Press, 1990). For an alternative interpretation, see Andrew W. Reddie et al., "Next-Generation Wargames," *Science* 362, no. 6421 (2018).
9. David Tyler Long, "Wargaming and the Education Gap: Why CyberWar 2025 Was Created," *The Cyber Defense Review* 5, no. 1 (2020): 185–200.
10. Andrew W. Reddie and Bethany L. Goldblum, "Evidence of the Unthinkable: Experimental Wargaming at the Nuclear Threshold," *Journal of Peace Research*, 2022.

11. Erik Lin-Greenberg, Reid B. C. Pauly, and Jacquelyn G. Schneider, "Wargaming for International Relations Research," *European Journal of International Relations* 28, no. 1 (2022): 83–109.

12. Kaja Silverman, *The Subject of Semiotics* (New York: Oxford University Press, 1989).

13. K. Salen and E. Zimmerman, *Rules of Play-Game Design Fundamentals* (Cambridge, MA: MIT Press, 2003); this is referred to as "consequence-based outcomes" by Erik Lin-Greenberg, Reid Pauly, and Jacquelyn Schneider, "Wargaming for International Relations Research," Social Science Research Network, 2020, SSRN ID 3676665.

14. Doing this work, we argue, is central to addressing "the technical and infrastructural perplexities" associated with cyberspace and wargaming; see Nina Kollars and Benjamin Schechter, "Pathologies of Obfuscation: Nobody Understands Cyber Operations or Wargaming," Atlantic Council, 2021, www.atlanticcouncil.org/in-depth-research-reports/issue-brief/pathologies-of-obfuscation-nobody-understands-cyber-operations-or-wargaming/.

15. Salen and Zimmerman, *Rules*.

16. Benjamin Schechter, Jacquelyn Schneider, and Rachael Shaffer, "Wargaming as a Methodology: The International Crisis Wargame and Experimental Wargaming," *Simulation & Gaming* (January 2021).

17. Reddie et al., "Next-Generation Wargames," 1362–64; Erik Lin-Greenberg, "Wargame of Drones: Remotely Piloted Aircraft and Crisis Escalation," Social Science Research Network, 2020, SSRN ID 3288988.

18. Benjamin Jensen and David Banks, "Cyber Operations in Conflict: Lessons from Analytic Wargames," University of California at Berkeley Center for Long-Term Cybersecurity, https://cltc.berkeley.edu/cyber-operations/.

19. This survey experiment also draws on conclusions drawn from *Netwar*, a game designed to collect the dynamics between a government, a violent nonstate actor, a major international firm, and a cyber activist network in the cyber domain.

20. Jensen and Valeriano, "Cyber Escalation Dynamics."

21. Jensen and Valeriano, 11.

22. Reddie et al., "Next-Generation Wargames," 1362–64.

23. Lucas Kello, "The Meaning of the Cyber Revolution: Perils to Theory and Statecraft," *International Security* 38, no. 2 (2013): 7–40; Buchanan and Cunningham, Preparing the Cyber Battlefield"; Buchanan, *Cybersecurity Dilemma*; Borghard and Lonergan, "Cyber Operations," 122–45.

24. James M. Acton, "Cyber Warfare & Inadvertent Escalation," *Dædalus* 149, no. 2 (2020): 133–49; Buchanan and Cunningham, "Preparing the Cyber Battlefield"; Richard J. Harknett and Max Smeets, "Cyber Campaigns and Strategic Outcomes," *Journal of Strategic Studies*, 2020, 1–34.

25. E.g., see Tiffany Bao et al., "How Shall We Play a Game? A Game-Theoretical Model for Cyber-Warfare Games," *Proceedings of IEEE 30th Computer Security Foundations Symposium*, 2017, 7–21. Or, for a survey of existing methods, see Yuan Wang et al., "A Survey of Game Theoretic Methods for Cyber Security," *Proceedings of IEEE First International Conference on Data Science in Cyberspace*, 2016, 631–36.

26. Igor Kotenko and Andrey Chechulin, "A Cyber Attack Modeling and Impact Assessment Framework," *Proceedings of 5th International Conference on Cyber Conflict*, 2013, 1–24; Noam Ben-Asher and Cleotilde Gonzalez, "Cyber War Game: A Paradigm for Understanding New Challenges of Cyber War," *Cyber Warfare 2015*, 207–20.

26. Scott de Marchi and Scott E. Page, "Agent-Based Models," *Annual Review of Political Science* 17, no. 1 (2014): 1–20.

27. Put in blunt terms, analysts should be wary of the "garbage in, garbage out" problem associated with any synthetic data-generating process.

28. Vladislav D. Veksler et al., "Cognitive Models in Cybersecurity: Learning from Expert Analysts and Predicting Attacker Behavior," *Frontiers in Psychology* 11 (2020).

29. Ryan Shandler, Michael L. Gross, and Daphna Canetti, "A Fragile Public Preference for Cyber Strikes: Evidence from Survey Experiments in the United States, United Kingdom, and Israel," *Contemporary Security Policy*, 2021, 1–28; Michael L. Gross, Daphna Canetti, and Dana R. Vashdi, "The Psychological Effects of Cyber Terrorism," *Bulletin of the Atomic Scientists* 72, no. 5 (2016): 284–91.

30. Sarah Kreps and Jacquelyn Schneider, "Escalation Firebreaks in the Cyber, Conventional, and Nuclear Domains: Moving Beyond Effects-Based Logics," *Journal of Cybersecurity* 5, no. 1 (2019).

31. Sarah E. Kreps and Debak Das, "Warring from the Virtual to the Real: Assessing the Public's Threshold for War on Cyber Security," January 14, 2017, http://dx.doi.org/10.2139/ssrn.2899423; Nadiya Kostyuk and Carly Wayne, "The Microfoundations of State Cybersecurity: Cyber Risk Perceptions and the Mass Public," *Journal of Global Security Studies*, 2020.

32. Jenny Oberholtzer et al., "Applying Wargames to Real-World Policies," *Science* 363, no. 6434 (2019): 1406.

33. Joshua D. Kertzer, "Re-Assessing Elite-Public Gaps in Political Behavior," *American Journal of Political Science*, 2020.

34. Floyd J. Fowler Jr. and Floyd J. Fowler, *Improving Survey Questions: Design and Evaluation* (New York: Sage, 1995).

35. Martin C. Libicki, "Expectations of Cyber Deterrence," *Strategic Studies Quarterly* 12, no. 4 (2018): 44–57; Will Goodman, "Cyber Deterrence: Tougher in Theory Than in Practice?" *Strategic Studies Quarterly* 4, no. 3 (2010): 102–35; Uri Tor, "'Cumulative Deterrence' as a New Paradigm for Cyber Deterrence," *Journal of Strategic Studies* 40, nos. 1–2 (2017): 92–117; Alex S. Wilner, "US Cyber Deterrence: Practice Guiding Theory," *Journal of Strategic Studies* 43, no. 2 (2020): 245–80; Aaron F. Brantly. "The Cyber Deterrence Problem," *Proceedings of International Conference on Cyber Conflict*, 2018, 31–54; M. D. Crosston, "World Gone Cyber MAD: How 'Mutually Assured Debilitation' Is the Best Hope for Cyber Deterrence," *Strategic Studies Quarterly* 5, no. 1 (2011): 100–116; Amir Lupovici, "The 'Attribution Problem' and the Social Construction of 'Violence': Taking Cyber Deterrence Literature a Step Forward," *International Studies Perspectives* 17, no. 3 (2016): 322–42; Alexander Klimburg, "Mixed Signals: A Flawed Approach to Cyber Deterrence," *Survival* 62, no. 1 (2020): 107–30.

36. Brendan Rittenhouse Green and Austin Long, "Conceal or Reveal? Managing Clandestine Military Capabilities in Peacetime Competition," *International Security* 44, no. 3 (2020): 48–83.

3

WARGAME RESEARCH on CYBER and NUCLEAR CRISIS DYNAMICS

BENJAMIN H. SCHECHTER, JACQUELYN SCHNEIDER, AND RACHAEL SHAFFER

The prosperity and well-being of nation-states, businesses, and individuals are now inextricably linked to cyberspace.[1] These dependencies are evident in the ubiquity of our smartphones, data-driven health care systems, and the vast supply chains that underpin the global economy. These dependencies are also why we study cybersecurity—to help safeguard people, organizations, and societies from cyber risks.[2]

Cybersecurity is a hard problem, no matter how it is scoped. It exists at the intersection of people, processes, and technology.[3] However, within cyber and cybersecurity research, there is an understandable tendency to focus more on the technology (e.g., hardware, software, and data) and, to an extent, the processes (e.g., security training and incident response), but less on the human dimension.[4] This deficit is particularly acute at the strategic level—the level at which corporate or national strategy is formulated and implemented. The deficit is partly driven by a lack of observational data on human decision-making during cyber crises.[5]

There are efforts to address this empirical shortfall.[6] Nevertheless, the problems are manifold. Data are often inaccessible (e.g., siloed in classified networks or corporate systems) and poorly structured. Additionally, gathering observational data about executive-level decision-making is challenging, since it requires access to a limited pool of senior decision-makers.[7] These problems are most acute for scholars doing generalizable, causal research (vs. descriptive or exploratory research).[8]

Despite all the challenges, it is possible to conduct good research and produce valid insight. As argued throughout this book, wargames are one tool for generating data to enable research and contribute to our understanding of cybersecurity.[9] This tool can generate data by bringing together people with notional processes and technologies, replicating the conditions of a conflict or crisis to study human decision-making. Yes, wargames abstract from the "real" world. Still, the decisions that players make in the game world can be representative of those they would make in a real crisis.[10]

The idea of using wargames to explore cyber-related issues is not new. As early as 1995, analysts at the RAND Corporation conducted a series of wargames to explore the emerging concept of "information warfare."[11] However, until the resurgence of wargaming in 2015 that we describe below, cyber wargames were a niche within what was already a relatively small and esoteric community of professional wargamers.[12] Today, cyber wargames are more common, but they are still not entirely mainstream within the defense establishment or among national and international security scholars.

This chapter explores cyber wargames for research, beginning with a brief history of cyber wargames until the contemporary period, where this book picks up. We also describe two recent examples of cyber wargames for research: the 2017 *Navy–Private Sector Critical Infrastructure Wargame* and the *International Crisis War Game*. The 2017 *Navy–Private Sector Critical Infrastructure Wargame* tested how different cyberattack characteristics—attackers, targets, and effects—affected the private sector and how it might or might not work with the US military. While instructive, this game was run only once. In partial contrast, the *International Crisis War Game* was an iterative experimental wargame series run between 2018 and 2020.[13] This game series tested the relationship between cyberspace operations and nuclear stability, answering important questions about cyber–nuclear crisis dynamics. Based on our lessons learned, this chapter concludes with a look at the potential of wargames for cyber research and future applications.

THE BRIEF HISTORY OF CYBER WARGAMES FOR RESEARCH

The history of wargaming is reasonably well documented, but there is no comprehensive history of cyber wargaming.[14] "Wargames are as old as civilization," according to Matthew Caffery, "and perhaps older."[15] Cyber wargames followed the creation of cyberspace, largely credited to a series of computer networks and information protocols developed by the Advanced Research Projects Agency during the 1960s and 1970s.[16] While some cyber wargames remain shrouded in classified mystery, at least four early examples stand out: *Day After . . . in Cyberspace*; *Eligible Receiver*; *Zenith Star*; and *Digital Pearl Harbor*. These games represent the opening intellectual salvos in cyber wargaming.

Day After . . . in Cyberspace

The 1991 Gulf War was heralded as "the first information war." Even though this concept was not new at the time, the US Department of Defense (DoD) had yet to fully understand or integrate the challenges and opportunities of digital information or cyberspace into military strategy or operational planning. In 1995, the DoD established an Information Warfare (IW) Executive Board, chaired by the deputy secretary of defense, "to provide a forum for the discussion and advancement of IW strategies, operations, and programs involving DoD."[17] The RAND Corporation was, in turn, tasked to conduct a set of exercises to frame and explore IW. Using its *"The Day After . . ."* methodology, RAND developed a series of exercises based on a notional conflict between Iran and the United States.[18]

Evolving versions of these exercises were conducted six times between January and June 1995.[19] Over the six iterations, participants included senior military officers, DoD civilians, and participants from the intelligence community, academia, and industry. The iterated approach enabled a gradual refinement of the strategic concepts and additional exploration of the national security implications of IW. The game helped identify several key features of IW: the low entry cost for actors in cyberspace, a blurring of traditional geographic borders, an expanded role for perception management, challenges for strategic intelligence, problems with tactical warning and attack assessment, the difficulty of building and sustaining coalitions in cyberspace, and finally the vulnerability of the US homeland.[20] For better or worse, most, if not all, these features remain relevant more than twenty-five years later.

Eligible Receiver

Joint Exercise Eligible Receiver was an annual future-focused wargame. In 1997, the National Security Agency (NSA) director, Lieutenant General Kenneth Minihan, was inspired to use *Eligible Receiver* to "test the vulnerability of the US military's networks to a cyberattack."[21] The resulting exercise pitted an NSA red team—playing the role of North Korean, Iranian, and Cuban forces—against an interagency blue team composed of defenders from the DoD, parts of the intelligence community, other government departments, and private-sector participants.[22]

Eligible Receiver was structured in three parts and ran over four days, although preparations began months in advance. The first part was a tabletop exercise. In the exercise, the red team attempted to extract political concessions from the blue team by launching a notional series of coordinated cyberattacks on critical infrastructures across the United States. In the second part of the wargame, the red team attacked DoD information systems, from phones to email. Somewhat surprisingly, this was not notional. The red team actually penetrated and attacked real DoD networks, marking their passing and causing disruptions (blurring the boundaries between a traditional wargame and penetration testing). The third part of the wargame involved notional kinetic attacks. These attacks included a fictitious hostage crisis supported by real-world intelligence the red team had gathered during earlier attacks on DoD networks.[23]

This wargame was a crushing defeat for the cyber defenders.[24] In a 2004 interview, John Hamre, the deputy secretary of defense, recalls that *"Eligible Receiver* changed a lot of our consciousness about the vulnerability of cyber— not completely through society, by any means—but certainly within the defense establishment. I think within the security establishment in general, we've got a much better appreciation."[25] Arguably, this cyber wargame was also an important milestone in the long road to establishing the US Cyber Command.[26]

Zenith Star

Zenith Star was run two years later, in October 1999. This wargame was led by the commander of the Pentagon's new Joint Task Force–Computer Network Defense, which had been established in the aftermath of *Eligible Receiver*.[27] It was intended to assess the implementation of lessons learned from *Eligible Receiver* and test the new task force. *Zenith Star* involved over fifty players from across the DoD, playing a similar scenario to *Eligible Receiver*. The game showed modest improvements, but it also highlighted persistent coordination issues and vulnerabilities in the national infrastructure.[28]

Digital Pearl Harbor

Digital Pearl Harbor was a wargame run at the US Naval War College in June 2002 to examine the threat of cyber terrorism.[29] The game designers sought "to determine and design the most damaging attack to the United States delivered largely or completely through cyber means" by a terrorist organization.[30] The game included over eighty private-sector players from the financial services, energy, and telecommunication (including Internet service providers) critical infrastructure sectors. Not surprisingly, the designers assumed that "attacking multiple industries is likely to cause greater devastation than just attacking one sector alone." [31] A red team devised a series of technically feasible cyberattacks against each sector, and these attacks, in turn, drove the wargame.[32]

The results indicated that a *"Digital Pearl Harbor"* type cyberattack against the United States was unlikely or, at least, had "limited possibility."[33] This wargame concluded that the hypothetical cyberattacks against critical infrastructure were not existential. Still, their disruption could adversely affect public well-being: "All industries except the Internet itself seem to have significant vulnerability to attack, but probably not to the extent of creating (singly or in combination) a *Digital Pearl Harbor*." [34] Only through substantial physical intervention—that is, a physical attack in conjunction with a cyberattack—were seriously harmful effects generated. The findings pointed to several important roles for government in coordination, crisis response, and restoring public confidence.[35] The game also shaped participants' perceptions, with 79 percent stating that they believed a significant cyberattack was likely within the next two years.[36]

These wargames laid the early foundation for research through cyber wargaming. They were primarily directed at explorative and descriptive research to understand the emerging—but not entirely new—intersections of information warfare, cyber operations, and national security. Over time, other research, educational, and training wargames were developed, such as the National Cyber Exercise or Cyber Storm, started in 2006 by the Department of Homeland Security and *GridEx*, as described in chapter 10.[37]

Another turning point came in 2015. The then–deputy secretary of defense, Bob Work, led a charge to reinvigorate the practice of wargaming with calls for more and improved wargaming across the DoD.[38] Wargames were viewed as a tool to assess the emerging technologies supporting the Third Offset Strategy, a controversial effort to develop and leverage technologies to better compete militarily with China and Russia.[39] Cyber wargaming, technology wargaming—like those focusing on artificial intelligence (AI) and autonomous systems—and conventional wargames all enjoyed a resurgence.[40] The next two games we discuss are from the post-2015 period.

CYBER IN WARGAMES

The 2017 *Navy–Private Sector Critical Infrastructure Wargame* (hereafter referred to as the *Critical Infrastructure Wargame*) and the *International Crisis War Game* (*ICWG*) are different examples of what might be called "modern" cyber wargames. The *Critical Infrastructure Wargame* focused on the relationship between the private sector and government, examining how cyberattacks might influence the public–private relationship. Players were drawn from the private sector and federal, state, and local governments. In a different game with different research questions, *ICWG* put players in the role of national leaders and asked how cyberspace operations affected the calculus of nuclear weapons use in the context of a crisis between nuclear-armed states. *ICWG* was run iteratively with a diverse range of participants, including senior decision-makers.

Both games are quasi-experimental, a specialized form of wargames that apply social science methods of treatments and control to test hypotheses about human decision-making.[41] A core element of these games is the focus on decreasing bias and deriving data collection methods that allow researchers to examine the effect of independent variables on outcomes.[42] These elements require experimental cyber wargames to frame "cyber" carefully, which, as described throughout this book, can be challenging.

The 2017 Navy–Private Sector Critical Infrastructure Wargame

The *Critical Infrastructure Wargame* was run in July 2017 at the US Naval War College. This game was developed to answer two questions: "When do cyberattacks reach the level of a national security incident?" And "when should the [DoD] be involved and in what capacity?"[43] The scenario focused on cyberattacks against critical infrastructure, a highly salient issue following a wave of major cyber incidents, including WannaCry, Petya, and NotPetya.[44] To capture the dynamic relationship between the private sector and government, the game incorporated players from most of the private critical infrastructure sectors identified by the Department of Homeland Security, along with participants from every level of government—all playing roles in a fictitious Blue State.[45]

The wargame ran two scenarios over two days, each with different notional antagonists. On the first day, the antagonist was a hostile country (Red State). On the second day, it was a violent extremist organization (Red Extremist Organization). The game was designed around scripted cyberattacks from hostile actors directed at critical infrastructure sectors. The attacks had predetermined effects, causing either virtual damage to networks and systems, physical damage to systems or cyber-physical systems, a loss of life, or a nuclear effect (an incident that generates a nuclear or similarly catastrophic effect). These scripted cyberattacks were the experimental treatment within the game; up to four scripted attacks targeted every sector in each scenario, one for each effect category (except when no viable effect was possible). The different cyberattack targets and effects were intended to assess which attacks players were most likely to interpret as having become a national security incident that required government intervention.[46]

The private-sector and government players had different actions available in the game based on their real-world capabilities. The action options were loosely constrained by existing institutions, processes, and the research objective. For example, the private sector could engage in internal corporate actions (e.g., internal incident response and calling an outside cybersecurity firm), coordinate actions across their critical infrastructure sector, and conduct strategic communications. Most crucially, they could make specified requests for assistance to the government—a mechanism designed to help measure and assess private industries' desired role for government intervention. The government players also had leeway in their actions, including acting against the aggressor and providing support to the private sector based on their requests.[47]

The representation of cyberspace in this game largely consisted of the scripted cyberattacks; no additional information was provided. Private-sector players, for example, were not given detailed information about their corporate networks; nor were the government players given detailed information about their offensive cyber capabilities. We designed the game this way to avoid adding unnecessary details beyond the scripted attacks that might inadvertently distract the players' attention from the relevant information without adding analytic value.

With this in mind, there were four principal design considerations for the scripted cyberattacks. First, because each sector only received a handful of cyberattacks and they had to have enough impact to motivate players to care— namely, they needed to be above the threshold of business as usual. Second, the effects of the cyberattacks were framed in a business context; why should corporate executives care? Third, most information players received had to be private, so players had to decide if and when to share information. Finally, the scripted cyberattacks had to have stop conditions; for example, a cyberattack was not viable if players took their network offline.

Our findings from this game indicated that the private sector tended to manage the technical aspects of a cyber crisis internally and that its threshold for requesting government assistance was high. However, private-sector players did identify an important role for the government in facilitating information sharing. Additionally, there was high demand for the government's emergency response and management resources when physical damage occurred, lives were lost, or there was a risk of widespread panic. Although these requests were mostly directed to state and local governments, players requested that the DoD help stop cyberattacks and prevent further attacks.[48]

The International Crisis War Game

The *ICWG* was developed to answer questions about the intersection of cybersecurity and a nuclear crisis, specifically, "how do cyber operations affect nuclear stability?"[49] More specifically, how do cyber vulnerabilities and exploits within nuclear command and control affect decisions to use nuclear weapons? Nuclear command, control, and communications systems (NC3) are the underlying infrastructure of modern strategic forces. They comprise the "facilities, equipment, communications, procedures, and personnel that enable presidential nuclear direction to be carried out."[50] Digital NC3 also depends on cyberspace. The wargame tested how cyberspace operations directed at disrupting NC3 and a state's ability to use its nuclear weapons affected decision-making about whether to use those weapons.

Wargames are particularly well suited to answer important research questions about the cyber-nuclear interface. First, real-world, empirical data on the nuclear and cyber intersection are scarce. Second, the core research question is about strategic decision-making and not the technical aspects of cyberspace or nuclear operations. Finally, there is a significant body of theoretical work on the effects of cyberspace operations on crisis stability and nuclear use to build upon. As with any experiment, a strong theoretical foundation is important to create an efficient, effective, and valid wargame.[51]

The *ICWG* had five to six players on a team assume roles in the cabinet of a notional country pitted against a scripted adversary state with no players. Both states had nuclear weapons and symmetrical military, economic, and diplomatic capabilities. The game was divided into two scenarios. The first scenario was a crisis over a contested region. In the second higher escalation scenario, the adversary state invaded the players' state. The second scenario was designed to simulate an existential threat, since the game needed a crisis of sufficient intensity to encourage players to discuss using their nuclear weapons seriously.[52]

Each team was assigned a condition consisting of two binary pieces of information—this was the experimental treatment. First, they either possessed the necessary network access to use a specialized cyber exploit to attack their adversary's NC3 or not. Second, they were either vulnerable to a cyberattack against their NC3 by the adversary or not. Thus, players got one of four potential experimental conditions. Some teams had both an access/exploit and a vulnerability; others had a mix or neither. Players were told that the NC3 attacks could disrupt the ability to use nuclear weapons for an unknown period.

The wargame was designed to examine how these experimental treatments affected two kinds of players' decisions: the decision to use or not use cyberspace operations against their adversary's NC3, and the decision to use or not use nuclear weapons. The cyber-related information players received needed to be easily comprehensible to participants with a wide range of cyber expertise. We eliminated unnecessary technical jargon to facilitate gameplay, and facilitators provided additional explanations only if appropriate. Additionally, the game did not prompt players to dwell on cyberspace operations' theoretical and practical challenges, such as attribution, capabilities attrition, or unintended consequences. Nevertheless, we preserved the realistic uncertainty of the success or failure of cyberspace operations. Players with the exploit were told that it had a "high probability of success," but they were offered no guarantees. The goal was to neither encourage nor discourage the use of cyberspace operations by design.[53]

To simulate a real crisis, players had access to all the tools of the state, including diplomatic, economic, and conventional military options. Cyberspace operations were abstracted to a similar strategic level as the other instruments of statecraft and separated into a stand-alone category among these other tools. Players were provided a set of target categories, such as "cyberattack on civilian target," "cyberattack on military target," and, for players with the access/exploit, "cyberattack on nuclear NC3." Players were also asked to provide written details about the targets they selected.

When the game series concluded, we had run the *ICWG* with 580 participants at 12 locations across the United States, internationally, and virtually.[54] Empirically, our iterated wargame experiment found limited evidence to support the hypothesis that cyber vulnerabilities create incentives for deliberate escalation to nuclear use. This finding has important implications for the nuclear taboo—the "normative inhibition against the first use of nuclear weapons," which appears intact.[55] However, we also observed that players were still willing to conduct cyberspace operations and it was the cyber capability that ultimately drove the most risky strategies, including nuclear alert and counterforce campaigns.

Wargames require players to make decisions in response to the information they receive. Both games provided players with a constrained set of cyber information, scoping down cyberspace to the specific context of the research question. In the *Critical Infrastructure Wargame*, players were told they were hit by a series of cyberattacks. The cyberattacks were not the research focus per se. The game was structured to assess how players responded to the attack and if their response did or did not involve the government. In the *ICWG*, players were given specific information about offensive cyber capabilities and vulnerabilities, but only in the context of a larger military crisis. While cyberspace was incorporated as an important means of managing a crisis that included a nuclear component, cyber was intentionally not elevated above other tools of statecraft. Both games answer important questions with critical implications for scholars and policymakers. They also illustrate how to ask different research questions and represent cyberspace accordingly in pursuit of valid answers.

RESEARCH IN CYBER WARGAMING: MORE AND DIFFERENT ARE BETTER

Cybersecurity remains a hard problem, from a practical perspective and for social scientific research. It is not intractable, however. The *Critical Infrastructure Wargame* and *ICWG* framed cyberspaces differently to answer substantially different questions using similar methods for cyber wargaming. The results testify to the diversity and utility of cyber wargames and the continuing maturation of this kind of research practice. We see many opportunities to leverage cyber wargaming to examine new problems and answer questions about human decision-making regarding information technologies. We also see several opportunities to refine cyber wargaming methods, say, by better integrating experimental and social science methods, and by improving how cyberspace is abstracted and explained.

Despite its militant trappings, cyber wargaming can be leveraged for research beyond national security. Yes, early research often centered on the DoD and its implications for US military operations. Still, there is a pressing need to answer broader questions about cybersecurity and cyberspace across government, industry, and society at large. For example, consider understanding and managing cyber risk. The widespread dependence on cyberspace means that individual users, organizations, and nation-states are exposed to various risks when operating in cyberspace. Cyber wargames provide one vehicle to study and manage these risks, helping answer questions about risk characteristics, mitigation methods, and cyber incident response. Beyond helping mitigate risk, cyber wargames can also identify opportunities for advantage, serving as a tool for strategy and planning in both the public and private sectors.

The methodology of cyber wargaming is another promising area for innovation and improvement, both in how games are developed and how they leverage experimental methods to provide analytic results. For instance, studying how players understand and respond to different abstractions and representations of cyberspace merits research. The findings could have useful implications for cyber wargaming methods and cybersecurity research more broadly. Similarly, studying different approaches to how information is presented to players in a wargame, how scenarios are structured, and even how data about players' behavior are recorded could also help improve the experimental conditions of cyber wargames. From where we stand, cyber wargames offer exciting opportunities to answer important policy questions and explore new research problems.

NOTES

1. The views expressed in this chapter are those of the authors alone and do not necessarily represent those of the US Navy, US Department of Defense, US government, or any other institution.
2. International Organization for Standardization, "ISO/IEC TS 27100:2020 Information Technology, Cybersecurity, Overview and Concepts," 2020, www.iso.org/obp/ui/#iso:std:iso-iec:ts:27100:ed-1:v1:en.
3. Horst W. J. Rittel and Melvin M. Webber, "Dilemmas in a General Theory of Planning," *Policy Sciences* 4, no. 2 (1973): 155–69; Chris Inglis, "Cyberspace—Making Some Sense of It All," *Journal of Information Warfare* 15, no. 2 (2016): 17–26; R. J. Harknett and E. O. Goldman, "The Search for Cyber Fundamentals," *Journal of Information Warfare* 15, no. 2 (2016): 81–88.
4. Shari Lawrence Pfleeger and Deanna D. Caputo, "Leveraging Behavioral Science to Mitigate Cyber Security Risk," *Computers & Security* 31, no. 4 (2012): 597–611; Calvin Nobles, "Botching Human Factors in Cybersecurity in Business Organizations," *Holistica: Journal of Business and Public Administration* 9, no. 3 (2018): 71–88.
5. Walter E. Cushen, "War Games and Operations Research," *Philosophy of Science* 22, no. 4 (1955): 309–20; Peter P. Perla and Darryl L. Branting, "Wargames, Exercises, and Analysis," Center for Naval Analyses, 1986.

6. Brandon Valeriano and Ryan C. Maness, *Cyber War versus Cyber Realities* (New York: Oxford University Press, 2015).

7. Federico Casano and Riccardo Colombo, "Wargaming: The Core of Cyber Training," *Next Generation CERTs* 54 (2019): 88; Ivanka Barzashka, "Wargaming: How to Turn Vogue into Science," *Bulletin of the Atomic Scientists*, March 15, 2019, https://thebulletin.org/2019/03/wargaming-how-to-turn-vogue-into-science/.

8. Elizabeth M. Bartels, "Building Better Games for National Security Policy Analysis" (PhD diss, Pardee RAND Graduate School, 2020), 44–51; David E. McNabb, *Research Methods for Political Science: Quantitative and Qualitative Methods* (New York: Routledge, 2015), 96–140.

9. Cushen, "War Games"; Perla and Branting, "Wargames."

10. Benjamin Schechter, Jacquelyn Schneider, and Rachael Shaffer, "Wargaming as a Methodology: The International Crisis Wargame and Experimental Wargaming," *Simulation & Gaming* 52, no. 4 (2021): 513–26; Erik Lin-Greenberg, Reid B. C. Pauly, and Jacquelyn G. Schneider, "Wargaming for International Relations Research," *European Journal of International Relations* 28, no. 1 (2022): 83–109.

11. Rodger C. Molander et al., *Strategic Information Warfare: A New Face of War* (Santa Monica, CA: RAND Corporation, 1996).

12. Benjamin Jensen, "Welcome to the Fight Club: Wargaming the Future," *War on the Rocks*, January 4, 2019, https://warontherocks.com/2019/01/welcome-to-fight-club-wargaming-the-future/.

13. Jacquelyn Schneider, Benjamin Schechter, and Rachael Shaffer, "Hacking Nuclear Stability: Wargaming Technology, Uncertainty, and Stability," forthcoming.

14. Matthew B. Caffrey, *On Wargaming: How Wargames Have Shaped History and How They May Shape the Future*, Naval War College Newport Paper 43 (Newport: Naval War College Press, 2019), 11; Roger C. Mason, "Wargaming: Its History and Future," *International Journal of Intelligence, Security, and Public Affairs* 20, no. 2 (2018): 77–101; Charles Homans, "War Games: A Short History," *Foreign Policy* 31 (2011); John Curry, "The History of Wargaming Project," in *Zones of Control: Perspectives on Wargaming*, edited by Pat Harrigan and Matthew G. Kirschenbaum (Cambridge, MA: MIT Press, 2016), 33–43.

15. Caffrey, *On Wargaming*, 11.

16. Janet Ellen Abbate, *From ARPANET to INTERNET: A History of ARPA-Sponsored Computer Networks, 1966–1988* (Philadelphia: University of Pennsylvania Press, 1994).

17. Mary M. Gillam, *Information Warfare: Combating the Threat in the 21st Century* (Montgomery, AL: Air Command and Staff College, 1997), 22.

18. Molander et al., *Strategic Information Warfare*.

19. Molander et al.

20. Molander et al.

21. Fred M. Kaplan, *Dark Territory: The Secret History of Cyber War* (New York: Simon & Schuster, 2016), 65.

22. Rebecca Slayton, "What Is a Cyber Warrior? The Emergence of US Military Cyber Expertise, 1967–2018," *Texas National Security Review*, Winter 2021; Michael Martelle, "Eligible Receiver 97: Seminal DOD Cyber Exercise Included Mock Terror Strikes and Hostage Simulations," National Security Archive, August 1, 2018, https://nsarchive.gwu.edu/briefing-book/cyber-vault/2018-08-01/eligible-receiver-97-seminal-dod-cyber-exercise-included-mock-terror-strikes-hostage-simulations.

23. Fred Kaplan, "The NSA Hacked into the US Military by Digging through Its Trash," *Slate*, March 7, 2016, https://slate.com/technology/2016/03/inside-the-nsas-shockingly-successful-simulated-hack-of-the-u-s-military.html.

24. Martelle, "Eligible Receiver 97."

25. John Hamre, "Cyber War!" *Frontline*, April 24, 2003, www.pbs.org/wgbh/pages/frontline/shows/cyberwar/interviews/hamre.html.

26. William Cohen, "The 20-Year Climb to an Elevated CyberCom," *Federal Computer Week*, October 9, 2017, https://fcw.com/articles/2017/10/09/comment-cohen-cyber.aspx?m=2.

27. Slayton, "What Is a Cyber Warrior?"; Bob Drogin, "US Scurries to Erect Cyber-Defenses," *Los Angeles Times*, October 31, 1999, www.latimes.com/archives/la-xpm-1999-oct-31-mn-28266-story.html; Gerald Burton and Richard Phares, "Zenith Star," *IA Newsletter*, 2000.

28. James Adams, "Virtual Defense," *Foreign Affairs* 80, no. 3 (2001): 98–112, doi: 10.2307/20050154; Drogin, "US Scurries."

29. F. Caldwell, R. Hunter, and J. Race, "'Digital Pearl Harbor' War Game Explores 'Cyberterrorism,'" *Gartner Research, E-17-6580*, 2002; Thomas C. Greene, "Mock Cyberwar Fails to End Mock Civilization," *Register*, August 30, 2002, www.theregister.com/2002/08/30/mock_cyberwar_fails_to_end/.

30. Lewis T. Booker Jr., "Changing the Lines in the Coloring Book," US Naval War College, 2004, https://apps.dtic.mil/sti/pdfs/ADA425910.pdf.

31. Caldwell, Hunter, and Race, "'Digital Pearl Harbor' War Game"; Eric Purchase and French Caldwell, "Digital Pearl Harbor: A Case Study in Industry Vulnerability to Cyber Attack," in *Guarding Your Business* (New York: Springer, 2004), 47–64; Booker, "Changing the Lines."

32. Greene, "Mock Cyberwar."

33. Clay Wilson, "Computer Attack and Cyber Terrorism: Vulnerabilities and Policy Issues for Congress" Congressional Research Service, 2005.

34. Booker, "Changing the Lines."

35. Caldwell, Hunter, and Race, "'Digital Pearl Harbor' War Game," 63.

36. Wilson, "Computer Attack."

37. Department of Homeland Security, National Cyber Security Division, "Cyber Storm After Action Final Report," September 12, 2006, www.cisa.gov/sites/default/files/publications/Cyber%20Storm%20I%20After%20Action%20Final%20Report.pdf.

38. Bob Work and Paul Selva, "Revitalizing Wargaming Is Necessary to Be Prepared for Future Wars," *War on the Rocks*, December 8, 2015, https://warontherocks.com/2015/12/revitalizing-wargaming-is-necessary-to-be-prepared-for-future-wars/; Bob Work, "Memo to Pentagon Leadership on Wargaming," Department of Defense, February 9, 2015, https://news.usni.org/2015/03/18/document-memo-to-pentagon-leadership-on-wargaming; Aggie Hirst, "Wargames Resurgent: The Hyperrealities of Military Gaming from Recruitment to Rehabilitation," *International Studies Quarterly*, 2022.

39. Paul Norwood and Benjamin Jensen, "Wargaming the Third Offset Strategy," *Joint Force Quarterly* 83, no. 4 (2016): 38; Gian Gentile et al., *A History of the Third Offset, 2014–2018* (Santa Monica, CA: RAND Corporation, 2021).

40. Norwood and Jensen, "Wargaming"; Gentile et al., *History*.

41. Lin-Greenberg, Pauly, and Schneider, "Wargaming"; Schechter, Schneider, and Shaffer, "Wargaming"; Andrew W. Reddie et al., "Next-Generation Wargames,"

Science 362, no. 6421 (December 21, 2018): 1362–64, doi:10.1126/science.aav2135; Bethany L. Goldblum, Andrew W. Reddie, and Jason C. Reinhardt, "Wargames as Experiments: The Project on Nuclear Gaming's SIGNAL Framework," *Bulletin of the Atomic Scientists*, May 29, 2019, https://thebulletin.org/2019/05/wargames-as-experiments-the-project-on-nuclear-gamings-signal-framework/.

42. Erik Lin-Greenberg, "Game of Drones: The Effect of Remote Warfighting Technology on Conflict Escalation (Evidence from Wargames)," 2018, https://dx.doi.org/10.2139/ssrn.3288988; Lin-Greenberg, Pauly, and Schneider, "Wargaming"; Schechter, Schneider, and Shaffer, "Wargaming"; Brandon G. Valeriano and Benjamin Jensen, "Wargaming for Social Science," 2021, http://dx.doi.org/10.2139/ssrn.3888395.

43. Jacquelyn Schneider, Benjamin Schechter, and Rachael Shaffer, "Navy-Private Sector Critical Infrastructure War Game 2017: Game Report," US Naval War College, 2017, www.hsdl.org/?view&did=810802.

44. Andy Greenberg, "The Untold Story of NotPetya, the Most Devastating Cyberattack in History," *Wired*, August 22, 2018, www.wired.com/story/notpetya-cyberattack-ukraine-russia-code-crashed-the-world/; Savita Mohurle and Manisha Patil, "A Brief Study of Wannacry Threat: Ransomware Attack 2017," *International Journal of Advanced Research in Computer Science* 8, no. 5 (2017): 1938–40.

45. Schneider, Schechter, and Shaffer, "Navy-Private Sector."

46. Schneider, Schechter, and Shaffer, "Navy-Private Sector."

47. Schneider, Schechter, and Shaffer, "Navy-Private Sector."

48. Jacquelyn G. Schneider, "Cyber Attacks on Critical Infrastructure: Insights from War Gaming," *War on the Rocks*, July 26, 2017; Schneider, Schechter, and Shaffer, "Navy-Private Sector Critical Infrastructure War Game."

49. Schechter, Schneider, and Shaffer, "Wargaming."

50. Office of the Deputy Assistant Secretary of Defense for Nuclear Matters, *Nuclear Matters Handbook 2020*, 2020, 231, www.acq.osd.mil/ncbdp/nm/nmhb/.

51. Schechter, Schneider, and Shaffer, "Wargaming."

52. Jacquelyn Schneider, Benjamin Schechter, and Rachael Shaffer, "A Lot of Cyber Fizzle but Not a Lot of Bang: Evidence about the Use of Cyber Operations from Wargames," *Journal of Global Security Studies* 7, no. 2 (2022), https://doi.org/10.1093/jogss/ogac005.

53. Schneider, Schechter, and Shaffer, "Cyber Fizzle."

54. Schneider, Schechter, and Shaffer, "Cyber Fizzle."

55. Nina Tannenwald, "The Nuclear Taboo: The United States and the Normative Basis of Nuclear Non-Use," *International Organization* 53, no. 3 (1999).

4

WARGAMING INTERNATIONAL AND DOMESTIC CRISES: *ISLAND INTERCEPT* AND *NETWAR*

DAVID E. BANKS AND BENJAMIN M. JENSEN

Cyber technologies change the character of war at the operational level as well as the practice of strategic competition in the twenty-first century.[1] At the international level, nation-states use cyber operations for espionage, propaganda, economic warfare, sabotage, shaping foreign public opinion and policies, and to disrupt opponents' political and military capabilities.[2] At the domestic level, some state and nonstate actors use cyber operations to surveil and control populations, support organized crime, and undermine political opponents.[3] Nonstate actors—including the Daesh/Islamic State, Anonymous, drug cartels, and organized criminals—also use the Internet for recruitment, to collect information, for blackmail, in distributed denial-of-service attacks, and for others ends.[4]

Despite all this activity, the long-term significance of cyber operations is unclear. How much will state and nonstate actors come to rely on cyber operations? What role will cyber operations play in major international and domestic crises? And what types of cyber strategies will different actors adopt?

These questions motivated our research. In 2016 we began investigating how cyber operations by the United States' rivals, such as Russia and the Islamic State, might interact with more traditional security dynamics such as deterrence and coercion. However, at that point in time, large-scale cyber operations—while potentially threatening—had not yet occurred. This absence of real-world data determined our decision to rely on wargames to generate synthetic data, as described in chapter 2. We created two wargames: *Island Intercept* and *Netwar.* Although cyber wargames for government date back to the 1990s, our research was among the first academic studies to use wargames in the literature on cyber conflict studies.[5] Using the novel data generated by our wargames, we identified three "strategy profiles" that actors engaged in cyber conflicts might adopt. These strategy profiles in turn informed a survey experiment to determine individuals' strategic attitudes toward cyber operations. In 2018, Berkeley's Center for Long-Term Cybersecurity published our initial report, *Cyber Operations in Conflict: Lessons from Analytic Wargaming.*[6]

"""

In this chapter, we explain our original research using cyber wargames. First, we outline some of the challenges and opportunities. Second, we discuss our research goals, specifically with regard to conflict escalation. Third, we discuss *Island Intercept* and *Netwar* in detail, drawing special attention to how cyber capabilities and events were represented, implemented, and adjudicated. We also discuss data collection. Finally, we briefly discuss how these games were used in the broader research design and outline our findings.

ANALYZING CYBER OPERATIONS THROUGH WARGAMING

The challenges of studying cyber operations—technological change, hype, secrecy, and the frequency of cyber events, to name just a few—are noted throughout this book and in the broader literature. In response to these challenges, scholars have often resorted to applying old concepts to the current problems, most notably those associated with nuclear deterrence. But, as many have noted, this practice risks theoretical stagnation. It may also miss or misrepresent many of the unique properties of cyber warfare.[7]

One issue of interest and importance is the potential effects of cyber operations on escalatory dynamics during crises. Prominent accounts of escalation typically attribute it to some combination of conflicting interests, shrinking time horizons, and incentives not to back down.[8] When combined, these factors create incentives to move up the escalatory ladder.[9] Yet for an actor to know whether they should back down or escalate, they need to understand the signals sent by their opponent.[10] There are good reasons to think that cyber operations alter these dynamics in meaningful ways.[11] Do cyber capabilities encourage or discourage escalation, for example, and what does general unfamiliarity with cyber operations imply for decision-making?

To help study and understand escalatory dynamics in the cyber domain, we turned to wargaming. Three reasons motivated this decision. First, while cyberattacks are not infrequent, at the time of our research there were few if any large international or domestic crises in which cyber operations had a meaningful impact. Wargames allowed us to generate synthetic data to investigate a low-frequency but potentially high-consequence phenomenon. Thus, these games were similar to those run during the Cold War to assess the likelihood of nuclear weapon use in the event of a war between the United States and the USSR, such as the 1982 *Ivy League Game* that examined responses to a first strike attack on Washington.[12]

Second, the competitive nature of wargames provided a way to understand more than just individuals' general attitudes but also their specific *strategic preferences* regarding cyber operations. In an effort to best their opponents, players are not seeking to discover some analytic truth; rather, they do what they can to win. As British field marshal Bernard Montgomery noted, "battle is not a one-sided affair. It is a case of action and reciprocal action repeated over and over as contestants seek to gain position and other advantages by which they may inflict the greatest possible damage upon their respective opponents."[13] Thus, competition likely makes data on players' attitudes and preferences more externally valid than the responses of individuals who have simply been asked what they think about cyber operations in a survey, for example.[14] They also overcome some of the abstraction and overrationalization of game theory, which struggles to account for the important role that cognitive shortcuts, stress, and fear of loss play in actual decision-making.

Third, the dynamic nature of multiturn wargames allows participants to iterate and adapt their strategies. Dynamic play gives greater insight into the flexibility or rigidity of their strategy preferences. Players may even develop strategies during gameplay that would have been unknowable beforehand. Wargames can thus act not only as experiments but also as venues for exploration and discovery.[15] For instance, when United States–versus–Japan wargames were first held at the US Naval War College during the interwar period (1919–41), it was assumed that big fleet battles would be decisive for US victory in a Pacific war. However, repeated play of this scenario revealed the more decisive island-hopping strategy that was ultimately adopted by the US Navy when war broke out in 1941.[16] We were similarly curious to discover what types of strategies might be typical and/or effective in a conflict between cyber-capable opponents, and whether common fears of a potential "cyber Pearl Harbor" were merited.[17]

We designed and ran two wargames to investigate the role of cyber operations in political conflicts. In *Island Intercept*, players took the role of the United States or China in an escalating crisis over Taiwan. In *Netwar*, players took the role of either a national government, a violent nonstate actor, a multinational firm, or an activist organization during an escalating domestic conflict. Each wargame differed in scenario design and implementation. Analyzing the data generated by these wargames helped us isolate strategic preferences and to extract three overarching "theories of victory" that transcended the different crisis contexts.[18] After we identified these theories of victory, we plugged them into a survey experiment using Amazon Web Services' Mechanical Turk. This method allowed us to study the strategic preferences of larger numbers of people (i.e., nearly 3,000 respondents). The wargames and survey are described next.

WARGAMING INTERNATIONAL CYBER OPERATIONS: *ISLAND INTERCEPT*

We designed both games as rigid *kriegsspiels*, meaning that they had precise rules about decisions and adjudication.[19] For such wargame results to be meaningful, it is necessary to design games in which (1) players understand their in-game environment sufficiently well that they can make moves, and (2) that their decisions can be recorded in ways that create useful data. To meet the first criterion, we designed rule sets and detailed scenario briefs that helped players clearly understand the game. For the second criterion, though players could make various political, economic, military, and cyber or information decisions, we set up most of these decisions to represent attempts to either coerce their adversary or to protect themselves against such coercion. Coercion is "the use of threatened force, and at times the limited use of actual force to back up the threat, to induce an adversary to change its behavior."[20] The framing was intentional to encourage players to play as aggressively as possible to further prompt acute strategic thinking. This meant designing games with clear win conditions, and that limited the adoption of cooperative strategies.

The choice to use rigid rule sets—rather than more open game designs— had trade-offs. On the one hand, because the players were not all subject matter experts, we wanted to avoid improbable or unrealistic decisions. Rigid *kriegsspiels* could prevent such decisions because players were constrained by what the design explicitly permitted. On the other hand, because these designs were an expression of our interpretation of the existing literature on cyber conflict, these games could not be used to "discover" novel systemic elements of this phenomenon. However, this choice made sense because our research focused on players' decision-making in cyber conflict as currently understood rather than determining cyber conflict's general dynamics. To put it another way, the games' model (i.e., rules) of how political, military, and cyber resources and actions interacted was ours; but the responses and reactions of individual players—including their willingness to escalate—remained theirs.[21]

Island Intercept is a two-player rigid *kriegsspiel* representing a military crisis between the United States and China in the South China Sea. Set in the near future, the scenario begins with an advanced Taiwanese cruiser being disabled by a cyberattack and the cruiser's subsequent capture by the Chinese Navy. The decision to begin the scenario with a successful cyberattack is intentional, priming players to consider cyber operations as part of their strategy.

Island Intercept takes place on a game board depicting the Western Pacific and Chinese littoral. As shown in figure 4.1, this region consists of six sea areas and eight land areas, the latter including Taiwan, Japan, and territories along China's east coast. Players receive individual maps on which only their assets are recorded. They also receive individual briefing documents that outline the game rules and player objectives. All materials were designed to heighten immersion. Players take the role of local commanders with the ability to deploy military, cyber, and political assets, as well as make political requests of superiors. Players not only influence each other with their actions; they can also take actions

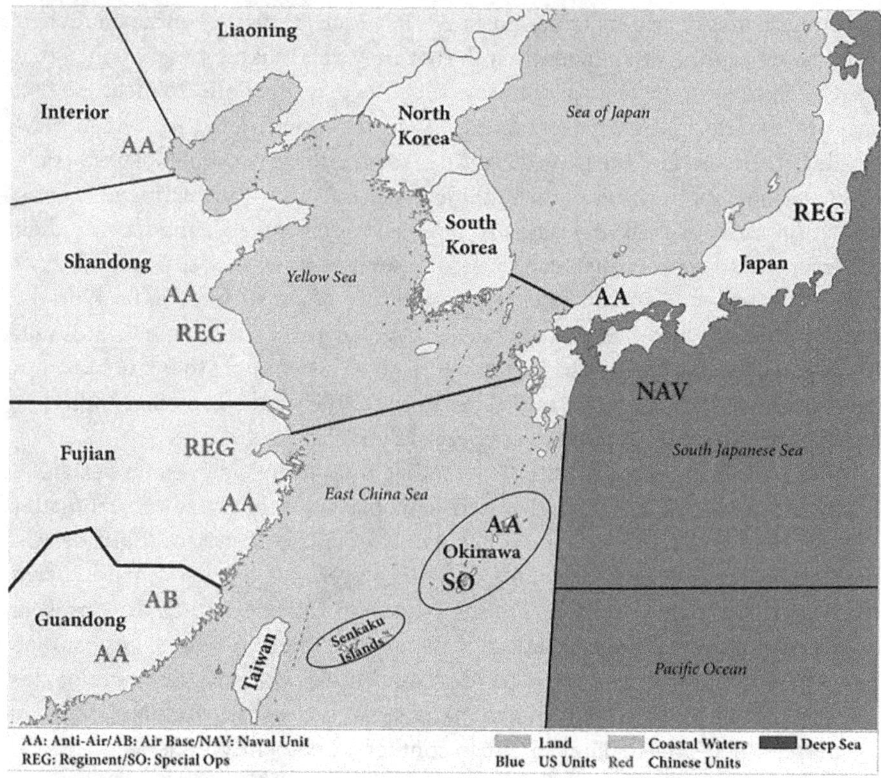

FIGURE 4.1. Controller Map from *Island Intercept*

that affect the political posture of Taiwan as well as influence how the United Nations responds to the crisis. Each turn represents one week. Gameplay continues for three turns.

This game was designed and then playtested over several iterations. In the first iteration, an initial playtest was run with two multiplayer teams. While a full record of the results was gathered, these data were not used in the final analysis. This iteration was run to help the designer identify potential bottlenecks and bugs in the game, after which the design was refined. It was then playtested again before being run in two sequences. In every instance, players participated in full debriefs, structured to include factual descriptions of the events as the players saw them.[22]

Representing Cyber Operations in Island Intercept

Each player team is given a list of potential cyber, military, and political actions they can take. To reduce potential "analysis paralysis" and increase immersion,

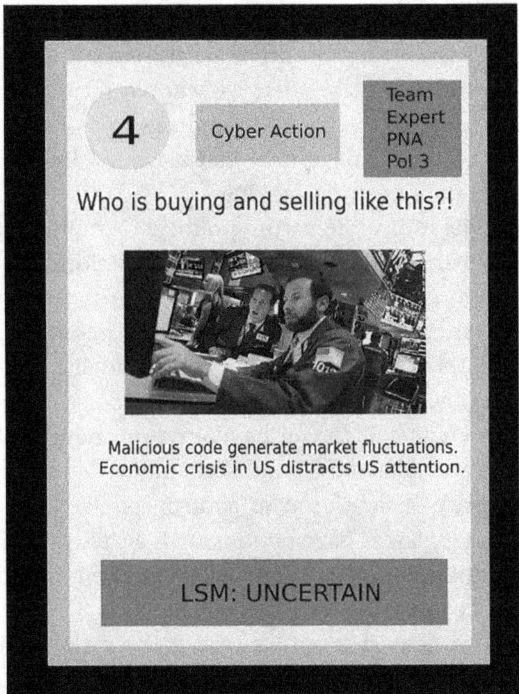

FIGURE 4.2. Example of Card from *Island Intercept*

each action is represented as a card, as illustrated in figure 4.2. This design feature was inspired by popular commercial games such as *Twilight Struggle* and *Labyrinth: The War on Terror*. Cards represent cyber, military, or political actions. Each card has a title (e.g., "Page not found!"), a picture, and an evocative description of the card's effects if it succeeds (e.g., "Opposing government embarrassed by the shutdown of its official websites"). The tops and bottoms of the cards record the action points, resources, and political permission levels required for the card to be selected. The bottoms of the cards record the action's likelihood of success.[23] Each turn, players have a limited number of action points, as well as limited resources to spend on cards. The resources required include specific assets as well as specific levels of "political permission" from superiors. This permission level is set at low at the beginning of the scenario, but it can change due to players requesting more permission or other events in the game.

In *Island Intercept*, the cyber domain is not represented as an independent space with its own unique topography or constraints. Rather, cyber operations are represented as actions that players can take, which might in turn affect activity on the game map or alter the behavior of other actors. Further, all three types of action available to players (political, military, cyber) are subject to the same action point / resources / political permission / "likelihood of success" system. We chose this design because we wanted participants to clearly and consistently understand the costs, risks, and rewards of all the decisions they can make. This increases the likelihood that players will understand the trade-offs associated with their decision-making.

Where cyber operations differ from political or military actions is in their in-game effects. Cyber actions threaten opponents along three classic dimensions of information security: confidentiality, integrity, and availability. Confidentiality refers to whether information can be kept private (e.g., via encryption and access controls). For cyberattacks on confidentiality, for example, players can attempt to gain access to their opponent's order sheet or deck of cards. Integrity refers to whether data can be altered or changed. For instance, the US player can attempt to create a wide-scale power blackout across China by attacking the integrity of data used to control its electrical grid. Availability refers to access. Here, players have options such as denying Global Positioning System access, creating "blue screens of death" on their opponent's computers, and denial-of-service attacks.

Players also have defensive options along these three dimensions to try to detect and/or prevent cyberattacks from succeeding. This range of cyber actions was designed to represent some of the major dynamics of cybersecurity discussed in the literature, including large-scale "doomsday" weapons, attribution problems, the potential to coerce opponents, and that cyber events might be responded to in the non–cyber domains.[24] While the Chinese and US players are given cards with similar effects, their capabilities in each domain are not identical. This design choice was made because the game emphasizes representing existing capabilities, not creating artificial "balance." Thus, players are not given the same number of cards; nor do the cards all have analogous effects. For instance, while the US player has a card that can cause a cyber doomsday event in China, the Chinese player does not have something similar. In total, the US player is able to choose from seven cyber actions, five political actions, and nine military actions. The Chinese player can choose from five cyber actions, seven political actions, and seven military actions.

Adjudication and Data Collection

Each turn, players select the cards they wish to play and then list them on an order sheet submitted to the game controller. For each card selected, players must confirm that they have sufficient action points and resources, and the requisite political permission to make these choices. Players are also asked to indicate whether there is a sequence in which they wish their actions to be attempted. The game controller then rolls two six-sided dice per action to determine the outcome of these orders on an "action resolution, effects, and detection" (ARED) table. These dice rolls determine (1) whether the action succeeds or fails, and (2) whether the opposing side detects its execution or attempted execution.

The ARED table provides the controller with prepared responses for every potential outcome of the players' actions. For instance, if the China player succeeds in a "Hunter" cyber action, then they receive this text from the controller at the end of their turn: "You have gained access to your opponent's network and have a record of all the orders given in the previous turn." If this action has been detected, the US player will, in turn, receive this text at the end of the turn: "You have detected attempts to hack all your networks." After all individual orders are processed for both players, the controller then rolls the dice to determine the activity of the United Nations and Taiwan in that turn. After all results for the turn are determined, the controller updates each player. The update informs players about changes to their map and the state of play. Then the next turn begins.

Of course, actions do not speak for themselves. To ensure that player motivations are not inferred incorrectly, they must also submit a form outlining the reasoning behind their actions for each turn. By asking players to explicitly discuss why decisions were made, we can determine whether options are avoided because they are seen as suboptimal or because they are not considered at all.[25]

The sequences we ran generated interesting findings. In the course of games, both sides (United States and China) were less aggressive than expected, including in their use of offensive cyber operations. For example, Chinese players frequently pursued a wait-and-see approach. This sometimes took the form of combining cyber espionage with traditional intelligence activities, such as military patrols and satellites and aerial reconnaissance. Other China players pursued a political path, lobbying at the international and regional levels, while assuming a defensive military posture and increasing their defensive cyber operations. Notably, US players typically adopted a cautious posture as well, tending to take diplomatic actions while positioning their military forces and seeking to gain advantage through cyber defense and espionage. Contrary to expectations in some of the literature, few players used paralyzing first strikes against mainland networks or used cyberattacks as the first stage of a larger conventional military attack.

WARGAMING DOMESTIC CYBER OPERATIONS: *NETWAR*

Netwar is a four-player rigid *kriegsspiel* that simulates multiparty competition in a domestic setting featuring four players: a government, a violent nonstate actor, a major international firm, and a cyber activist network. The game represents elements of the complex interactions governments face when confronting violent nonstate actors while also dealing with international firms and transnational advocacy networks that use cyber to coerce state and nonstate actors. The government player represents a middle-income country trying to suppress a violent extremist organization along its border. The violent nonstate player represents a transnational criminal network. The international firm player represents a *Fortune* 500 multinational company. The activist player represents a transnational advocacy group analogous to Anonymous.

Players have asymmetric capabilities. Some players are more capable in some domains than others (e.g., not all actors can engage in physical attacks in the security domain). Over the course of an in-game year, each player in *Netwar* competes for advantage in four domains: security, political, economic, and cyberspace. Gameplay continues for four turns, where each turn represents three months. The player with the greatest advantage in the most domains at the end of the fourth turn is declared the winner.

Each game starts with a briefing that describes the world, actors, and objectives to immerse players. From there on, each player is oriented to their movement sheet: essentially, a table listing actions players can take based on their available resources. Of note, the game features imperfect information, in that players only have partial information regarding what other players can do, as discussed in chapter 5. Players are welcome to communicate with one another, but they cannot share their movement sheets. This restriction preserves imperfect information and creates the possibility of credible commitment issues consistent with bargaining theory.[26]

Representing Cyber Operations in Netwar

The design of *Netwar* was inspired by *Chaturaji*—an ancient Indian game similar to chess. The game seeks to model coordination and competition challenges among varying interests by presenting multiple players with opportunities to contest multiple spaces simultaneously. This helps capture the confusing multidomain nature of cyber operations in the real world, where traditional political tools are combined with cyber capabilities in creative ways, such as Russia's use of information and influence operations, in which cyber espionage occurs alongside propaganda to undermine the confidence in democratic institutions.[27]

Each of the four domains privileges different actors and actions. The security domain is where the state and violent nonstate actor predominantly seek the initiative through activities ranging from targeting high-value individuals with drone strikes or raids to terrorist attacks and seizing strategic villages. Other actors seek to avoid physical threats and survive. In the political and economic domains, all four actors compete to mobilize support for their cause and undermine their opponents through a mix of lawfare, sanctions, illicit networks, social media, diplomacy, and propaganda. In the cyber domain, all four actors conduct offense, defense, and espionage to gain a position of relative advantage. Specific attention was paid to highlight the unique moves each player could make for cyber advantage. For example, the activists could employ a wiki-leak like data drop or doxing of incriminating evidence to undermine other players in the cyber domain.

Thus, cyber was represented as both a distinct domain of activity as well as a set of actions that players could undertake. However, all domains were abstractly represented, with no topographic differences between them. While each actor had different strengths and weaknesses in each domain, the types of trade-offs players needed to evaluate were the same across all four. This was an intentional design choice. We wished to ensure that players could readily understand the trade-offs associated with various decision options they had. Further, in keeping with literature on the difficulty of attribution, this game featured fog and friction in the cyber, political, and economic domains. Assigned moves were not guaranteed to occur; nor did players know others' capabilities or the moves they selected for each turn.[28]

This wargame was played with students at American University in the School of International Service and at the Marine Corps University with military officers ranging from captains to lieutenant colonels. Several interesting cyber dynamics emerged, consistent with cyber competition in the real world. First, players demonstrated a preference for cyber espionage to gain an information advantage. Because the game modeled imperfect information, players often sought to understand what the players on the other side were doing to hone their own strategies. Second, players showed a preference for misinformation, using false injects via cyber to take advantage of information asymmetries. Even though every player could employ cyber defenses, most preferred to engage in political warfare by leaking real and false information to alter others' strategic preferences. This dynamic also was used more frequently than cyber offense, with more deliberate offensive actions tending to be less opaque and designed to signal adversaries. This finding echoes scholarship on cyber signaling and findings from the 2020 US Cyberspace Solarium Commission.[29]

Adjudication and Data Collection

As with *Island Intercept,* the game controller in *Netwar* collects movement sheets, calculates results, and communicates these results back to the players. The controller counts offensive and defensive points based on players' moves to determine which players win a domain in each turn. Data from *Netwar* were also used to construct survey experiments and larger assessments about strategy preferences.

INTEGRATING WARGAME RESULTS INTO A SURVEY EXPERIMENT

Due to sample size and the differences in the game designs, the results of both *Island Intercept* and *Netwar* were too small to determine general attitudes toward cyber operations. However, players' decisions in these games were crucial for identifying "strategy profiles," which we then used to design a wider survey experiment.

In-game decisions and postgame rationales were studied to inductively determine the players' strategies regarding cyber operations. A number of specific questions guided this analysis. For example, did the players have a clear strategy throughout or did they work it out during the game? Was the strategy long term or short term? How cautious or aggressive was gameplay? And was the preferred mode of action military, political, economic, or cyber?

These player tendencies were then analyzed through the lens of principles of war. Principles of war originated during the Enlightenment as an analytical tool to help military professionals assess battles and campaigns.[30] Current US joint doctrine lists twelve principles: objective, offensive, mass, maneuver, economy of force, unity of command, security, surprise, simplicity, restraint, perseverance, and legitimacy.[31] Principles of war are an analytical framework that provides comparative logic that we used to differentiate different theories of victory.[32] Beyond just coding events as offensive or defensive, principles of war articulate a strategic hypothesis. Did the players assess that massing their political, economic, and cyber activities would achieve a decisive outcome? Did players predict that protecting their critical capabilities while harassing their opponents in multiple spheres would provide initiative? Five key principles were used to evaluate player strategies, as seen in table 4.1.

Based on these five principles, we discovered three general cyber strategy profiles that players used in both *Island Intercept* and *Netwar*:

- Mass and objective. *Use an escalatory cyber offensive to create a fait accompli.* Strike first, using the ambiguity of cyberspace to undermine your rival.
- Maneuver and surprise. *Cross-domain escalation and brinksmanship.* Escalate in another domain (military show of force, economic sanction threat, diplomatic threat) to pressure your rival to sue for peace, but wait to act in the cyber domain.
- Economy of force and security. *Test your opponent and limit escalation.* Probe your adversary with low-level cyber intrusions to signal resolve, but only retaliate at a higher level in cyberspace in another domain if they strike first (tit-for-tat). Harden your networks (cyber defense).

TABLE 4.1. Principles of War

Principle	Observation
Economy of force/effort	Take actions in multiple spheres along multiple lines of effort as opposed to massing effects to coerce an adversary.
Maneuver	Seek mismatches in exchanges that place the adversary in a position of disadvantage.
Mass / concentration of force	Coordinate mass coercive efforts in time and space to achieve a decisive result.
Objective/selection and maintenance of the aim	Direct coercive efforts toward a clearly defined, decisive point.
Security	Protect critical capabilities to produce future advantages.
Surprise	Seek shock and dislocation as a means of gaining advantage.

We used questions about these strategies as the basis of survey experiments conducted using Mechanical Turk. In the first survey ($n = 1,600$), respondents were asked to imagine themselves in the role of a great power in an emerging state-to-state crisis involving cyber operations—a scenario similar to *Island Intercept*.[33] In the second survey ($n = 1,600$), participants were also asked to imagine themselves as either state or nonstate actors engaged in an intrastate conflict—a scenario similar to *Netwar*.[34]

Combined with our wargame results, these survey experiments suggested a number of notable findings. In *Island Intercept* and the associated survey, we found that most participants were reluctant to use cyber operations in an escalatory manner. Instead, the majority choose to deescalate in the cyber domain. In *Netwar* and its attendant surveys, the results were more mixed. Only a plurality chose to deescalate in the cyber domain (a proclivity held more by people playing the role of democracies and less by autocracies).

Overall, our findings suggest that cyber weapons may be far less destabilizing than many commentators and policymakers typically assume. Actors were more likely to use military, economic, or diplomatic options before deciding to escalate in the cyber domain. Subsequent events suggest that these findings are robust. At the time in which we began our research and designed our games (2016–17), there was a growing consensus that the increase in cyber capabilities presaged a new era of "hybrid war" with more conflicts that occurred below the threshold of full conventional war, and in which cyberattacks would play a central role.[35] However, Russia's 2022 invasion of Ukraine belied those expectations. Rather than being a core feature of Russian offensive strategy, cyber has played a secondary role, much as it did in our games.[36]

Critics might contend that our representations of cyber operations and the cyber domain were oversimplifications, if not outright distortions. If this research aimed to analyze the outcomes of cyber operations, then this criticism would have merit. However, since we were interested in peoples' decision-making about cyber operations rather than the outcomes of such conflicts, we have more confidence about the external validity of our results. Nonetheless, we are conscious that what we analyzed are strategic preferences regarding cyber operations as we modeled them, not as they actually exist. While this condition limits the scope of our claims, it does not undermine our central finding: that individuals are reluctant to escalate in the cyber domain. This finding is supported by both our wargames and survey experiments, illustrating the original—and perhaps counterintuitive—contribution of our novel approach to social scientific research on cyber conflict.

NOTES

1. Brandon Valeriano, Benjamin M. Jensen, and Ryan C. Maness, *Cyber Strategy: The Evolving Character of Power and Coercion* (New York: Oxford University Press, 2020).

2. Austin Carson, "Analysis: Obama Used Covert Retaliation in Response to Russian Election Meddling; Here's Why," *Washington Post*, June 29, 2017, www.washington post.com/news/monkey-cage/wp/2017/06/29/obama-used-covert-retaliation-in -response-to-russian-election-meddling-heres-why/; Benjamin M. Jensen, "The Cyber Character of Political Warfare," *Brown Journal of World Affairs* 24, no. 1 (2017).

3. Mathew Burrows and Maria J. Stephan, *Is Authoritarianism Staging a Comeback?* (Washington, DC: Atlantic Council, 2015); Jennifer Windsor, Jeffrey Gedmin, and Libby Lu, "Authoritarianism's New Wave," *Foreign Policy*, June 23, 2009, http:// foreignpolicy.com/2009/06/03/authoritarianisms-new-wave/.

4. Ashley Fantz, "As ISIS Threats Online Persist, Military Families Rethink Online Lives," CNN, March 23, 2015, www.cnn.com/2015/03/23/us/online-threat-isis-us -troops/index.html; E. T. Brooking, "Anonymous vs. the Islamic State," *Foreign Policy*, November 13, 2015, http://foreignpolicy.com/2015/11/13/anonymous-hackers -islamic-state-isis-chan-online-war/; Elizabeth Flock, "Anonymous Cancels Operations against Drug Cartel, Says Kidnapped Member Has Been Found," *Washington Post*, November 4, 2011, www.washingtonpost.com/blogs/blogpost/post /anonymous-cancels-operations-against-drug-cartel-say-kidnapped-member-has -been-found/2011/11/04/gIQA11SzmM_blog.html?utm_term=.e53568290c50.

5. For other contemporary work, see Jacquelyn Schneider, "Cyber and Crisis Escalation: Insights from Wargaming," USASOC Futures Forum, 2017, https://paxsims

.files.wordpress.com/2017/01/paper-cyber-and-crisis-escalation-insights-from -wargaming-schneider.pdf; Reid B. C. Pauly, "Would US Leaders Push the Button? Wargames and the Sources of Nuclear Restraint," *International Security* 43, no. 2 (2018). Later studies include that by Erik Lin-Greenberg, "Wargame of Drones: Remotely Piloted Aircraft and Crisis Escalation," SSRN paper, August 22, 2020. For subsequent research covering cyber warfare, see Yuna Huh Wong et al., *Deterrence in the Age of Thinking Machines* (Santa Monica, CA: RAND Corporation, 2020); Stacie L. Pettyjohn and Becca Wasser, *Competing in the Gray Zone: Russian Tactics and Western Responses* (Santa Monica, CA: RAND Corporation, 2019); and Benjamin Schechter, Jacquelyn Schneider, and Rachael Shaffer, "Wargaming as a Methodology: The International Crisis Wargame and Experimental Wargaming," *Simulation and Gaming* 52, no. 4 (2021).

6. Benjamin Jensen and David Banks, *Cyber Operations in Conflict: Lessons from Analytic Wargames* (Berkeley: UC Berkeley Center for Long-Term Cybersecurity, 2018), https://cltc.berkeley.edu/2018/04/16/cyber-operations-conflict-lessons-analytic -wargames/. This research was funded by the Hewlett Foundation.

7. Erica D. Borghard and Shawn W. Lonergan, "The Logic of Coercion in Cyberspace," *Security Studies* 26, no. 3 (2017); Lucas Kello, "The Meaning of the Cyber Revolution: Perils to Theory and Statecraft," *International Security* 38, no. 2 (2013); Brandon Valeriano and Ryan C. Maness, *Cyber War versus Cyber Realities: Cyber Conflict in the International System* (New York: Oxford University Press, 2015); Erik Gartzke, "The Myth of Cyberwar: Bringing War in Cyberspace Back Down to Earth," *International Security* 38, no. 2 (2013); Thomas Rid, "Cyberwar and Peace: Hacking Can Reduce Real-World Violence," *Foreign Affairs* 92, no. 6 (2013); Robert M. Lee and Thomas Rid, "OMG Cyber! Thirteen Reasons Why Hype Makes for Bad Policy," *RUSI Journal* 159, no. 5 (2014).

8. Lisa J. Carlson, "Escalation, Negotiation, and Crisis Type," *Escalation and Negotiation in International Conflicts*, edited by I. William Zartman and Guy Olivier Faure (New York: Cambridge University Press, 2005), 215; George H. Quester, "Crises and the Unexpected," *Journal of Interdisciplinary History* 18, no. 4 (1988): 702; David E. Banks, "The Diplomatic Presentation of the State in International Crises: Diplomatic Collaboration during the US-Iran Hostage Crisis," *International Studies Quarterly* 63, no. 4, 2019.

9. Michael Brecher, "Crisis Escalation: Model and Findings," *International Political Science Review* (*Revue Internationale de Science Politique*) 17, no. 2 (1996); I. William Zartman and Guy Olivier Faure, eds., *Escalation and Negotiation in International Conflicts* (New York: Cambridge University Press, 2005), 4–6.

10. Thomas C. Schelling, *Arms and Influence* (New Haven, CT: Yale University Press, 1966), 116–18.

11. Jason Healey and Robert Jervis, "The Escalation Inversion and Other Oddities of Situational Cyber Stability," 2020, https://doi.org/10.26153/tsw/10962; Martin Libicki, *Crisis and Escalation in Cyberspace* (Santa Monica, CA: RAND Corporation, 2012); Martin C. Libicki and Olesya Tkacheva, "Cyberspace Escalation: Ladders or Lattices?" in *Cyber Threats and NATO 2030: Horizon Scanning and Analysis* (Brussels: NATO Cooperative Cyber Defence Centre of Excellence, 2020), 60; Benjamin Jensen and Brandon Valeriano, *What Do We Know about Cyber Escalation? Observations from Simulations and Surveys* (Washington, DC: Atlantic Council, 2019);

Erica D. Borghard and Shawn W. Lonergan, "Cyber Operations as Imperfect Tools of Escalation," *Strategic Studies Quarterly* 13, no. 3 (2019): 122–45.

12. T. B. Allen, "The Evolution of Wargaming: From Chessboard to Marine Doom," in *War and Games*, edited by T. J. Cornell and T. B. Allen (New York: Boydell Press, 2002), 243.

13. Nick Skellon, *Corporate Combat: The Art of Market Warfare on the Business Battlefield* (Boston: Nicholas Brealey, 1999), 4.

14. Erik Lin-Greenberg, Reid B. C. Pauly, and Jacquelyn G. Schneider, "Wargaming for International Relations Research," *European Journal of International Relations* 28, no. 1 (2022): 83–109; Schechter, Schneider, and Shaffer, "Wargaming."

15. Robert C. Rubel, "The Epistemology of War Gaming," *Naval War College Review* 59, no. 2 (2006): 22.

16. Peter P. Perla, *The Art of Wargaming: A Guide for Professionals and Hobbyists* (Annapolis: Naval Institute Press, 1990), 73.

17. Peter W. Singer and Allan Friedman, *Cybersecurity: What Everyone Needs to Know* (New York: Oxford University Press, 2014).

18. For more on wargaming and the "cycle of research," see Perla, *Art of Wargaming*, 287; and Peter Perla and E. D. McGrady, "Why Wargaming Works," *Naval War College Review* 64, no. 3 (2018).

19. For more on the different types of wargames, see Phillip E. Pournelle, "Designing Wargames for the Analytic Purpose," *Phalanx* 50, no. 2 (2017): 48–53.

20. Daniel Byman and Matthew Waxman, *The Dynamics of Coercion: American Foreign Policy and the Limits of Military Might* (New York: Cambridge University Press, 2002). There is a long tradition of studying coercion in international relations and military theory. For other definitions, see Schelling, *Arms and Influence*, 3; Carl von Clausewitz, *On War*, translated by Michael Howard and Peter Paret (Princeton, NJ: Princeton University Press, 1976), 75; and Robert Anthony Pape, *Bombing to Win: Air Power and Coercion in War* (Ithaca, NY: Cornell University Press, 1996), 4.

21. Elizabeth M. Bartels, "Building Better Games for National Security Policy Analysis: Towards a Social Scientific Approach" (Ph.D. diss., Pardee RAND Graduate School, 2020).

22. Guillaume Der Sahakian et al., "Setting Conditions for Productive Debriefing," *Simulation & Gaming* 46, no. 2 (2015): 197–208.

23. The likelihood of success can be highly likely, likely, uncertain, unlikely, or highly unlikely. Participants were not given the exact odds associated with these options.

24. Dale Peterson, "Offensive Cyber Weapons: Construction, Development, and Employment," *Journal of Strategic Studies* 36, no. 1 (2013): 120–24; Kello, "Meaning"; Rid, "Cyberwar"; Charles Glaser, *Deterrence of Cyber Attacks and US National Security*, GW-CSPRI-2011-5 (Washington, DC: Cyber Security Policy and Research Institute, 2011).

25. Phillip Sabin, *Simulating War: Studying Conflict through Simulation Games* (London: Continuum International, 2012), 59.

26. Robert Powell, "Bargaining Theory and International Conflict," *Annual Review of Political Science* 5, no. 1 (2002): 1–30; James D. Fearon, "Rationalist Explanations for War," *International Organization* 49, no. 3 (1995): 379–414.

27. Robert S. Mueller III, *Report on the Investigation of Russian Interference in the 2016 Election* (Washington, DC: US Department of Justice, 2019).

28. Kello, "Meaning"; Rid "Cyberwar."

29. Brandon Valeriano and Benjamin Jensen, "Building a National Cyber Strategy: The Process and Implications of the Cyberspace Solarium Commission Report," in *Proceedings of 13th International Conference on Cyber Conflict (CyCon)*, 2021, 189–214, https://ccdcoe.org/uploads/2021/05/CyCon_2021_book_Small.pdf.

30. Azar Gat, *A History of Military Thought: From the Enlightenment to the Cold War* (New York: Oxford University Press, 2002).

31. Chairman of the Joint Chiefs of Staff, JP 3-0, *Operations* (Arlington, VA: US Department of Defense, 2017).

32. On theories of victory, see Benjamin Jensen, *Forging the Sword: Doctrinal Innovation in the US Army* (Stanford, CA: Stanford University Press, 2016).

33. Four variations of the question were asked, to assess differences in behavior due to variations in relative power or issue salience.

34. Eight versions of this dispute were presented, varying on player role (state vs. nonstate), regime type (democratic vs. autocratic), and dispute type (ideological/religious vs. ethnic/minority rights).

35. Mark Galeotti, "Hybrid, Ambiguous, and Non-Linear? How New Is Russia's 'New Way of War'? *Small Wars & Insurgencies* 27, no. 2 (2016): 282–301.

36. Kari Paul, "'Catastrophic' Cyberwar between Ukraine and Russia Hasn't Happened (Yet), Experts Say," *Guardian*, March 9, 2022, www.theguardian.com/technology/2022/mar/09/catastrophic-cyber-war-ukraine-russia-hasnt-happened-yet-experts-say.

5

IMPERFECT INFORMATION IN CONVENTIONAL WARGAMING

CHRIS DOUGHERTY AND ED MCGRADY

Imagine a game of poker where the players can see each other's hands. It would hardly be considered a game. Instead, the tense competition of probability and psychology that people find so riveting would be replaced with a tedious exercise of odds and arithmetic. This is the problem of perfect information in games. All the interesting dynamics of the game—from betting and raising to bluffing and folding—are driven by *im*perfect information about what cards other players hold.

Unfortunately, when it comes to wargaming in the US Department of Defense (DoD), we often observe a pitiful version of poker—with all the cards face up. Players on opposing teams typically gaze at a game map that shows most if not all the military forces in play for both sides. This "God's-eye view" provides players with perfect information about themselves and their adversaries. Granted, some wargames allow for secret moves, and most hide submarines from the opposing team to simulate the opacity of the undersea domain. All too often, however, both sides get to see the other's "cards." Philip Sabin, author of the wargaming book *Simulating War*, pejoratively calls these games "chess with dice."[1]

We have long been frustrated by this unrealistic aspect of defense wargaming. Chris's frustration crystallized when he started playing "block-style" games for entertainment. Block-style games are named for the wooden blocks they use to represent military forces. The blocks stand on their edge with the military unit represented on one side and a blank face on the other. The games he played—Columbia Games' *Napoleon* and GMT Games' *Triumph and Tragedy*—were simple representations of combat.[2] Nevertheless, the blocks allowed for the fog of war and deception, since setting the blocks on edge prevents other players from seeing what kind of force the game piece represents.

This small addition of uncertainty was significant. In contrast to the wargames Chris observed during his day job at the DoD, with block games, players did not know the type or quality of their opposition until the fighting commenced. As a result of this uncertainty, players tended to be more conservative, taking more time to scout enemy forces and dedicating more forces to each move as a hedge against the possibility that their opponent was stronger than they seemed. Ironically, these entertainment games also placed a greater cognitive demand on hobbyists than some serious wargames placed on professionals. Rather than simply looking at a shared map, players diligently watched their opponent's every move in the hope of gleaning clues about the forces in play.

The more Chris played block games, the more he realized how much harder and potentially more realistic these games could be for military strategists and operational planners in the Pentagon. When he left the Pentagon and joined the Center for a New American Security, Chris carried this bee in his bonnet. In 2019, he started working on a project to explore imperfect information in wargaming. Shortly thereafter, the global COVID-19 pandemic halted in-person events and forced the adoption of remote wargaming, with participants in separate locations communicating online through chat, streaming, and file sharing. Given this confluence of events, we began pushing the informational aspect of wargame designs in new and more realistic directions.

In this chapter, we argue that the role of information—specifically imperfect information and the means to degrade information—is foundational to any realistic wargame. Imperfect information has always been important in real life. This fact should be evident, even without playing Napoleonic games with wooden blocks or, as is common among students in professional military education, quoting Sun Tzu.[3] Imperfect information is even more important in modern and future warfare, given the reliance on information for key missions like long-range precision attacks, and the impact of cyber operations therein.

THE PROBLEMS WITH PERFECTION

As Carl von Clausewitz explains in *On War*, "the first, the supreme, the most far-reaching act of judgment that the statesman and commander have to make is to establish by that test the kind of war on which they are embarking; neither mistaking it for, nor trying to turn it into, something that is alien to its nature."[4] Just as knowledge of other players' cards changes the character of poker, perfect information in wargaming alters the character of the military competition, coercion, and conflict that professional wargames purport to represent. In the real world, military commanders (1) do not know precisely where their opponent is, (2) do not know precisely where friendly forces are, (3) do not know what forces their opponent has, (4) do not know their opponents' intentions, and (5) do not know how fog and friction will interfere with their plans. In wargames, perfect information essentially eliminates these concerns for the players. Even adversaries' intentions—while not wholly transparent—are much easier to deduce when their disposition is known.

Giving players perfect information removes fog, friction, and the real possibility of deception in ways that run counter to even a cursory understanding of military history. Perfect information games create an illusion of precision and certainty in which warfare looks more like a mathematical exercise in resource allocation rather than a struggle of competing intellects and wills within a maelstrom of difficulty and poor information. These games boil down to which side can bring greater firepower to bear at longer ranges—becoming, essentially, mathematics problems answerable by calculating the effect of firepower (e.g., Lanchester square equations).[5]

Perfect information games also discount the importance of intelligence and communication in warfare. Countless battles and campaigns have turned on what commanders did or did not know about enemy and friendly forces, and whether they could communicate with their subordinates at critical moments. By making information "free," perfect-information wargames systematically undervalue organizations, systems, and operations designed to gather, process, and transmit it, as well as efforts to exploit, disrupt, and manipulate information.

Information and information systems will likely be even more important in the future given the widespread use of digital information technologies by modern military forces for intelligence, command and control (including communication), and precision targeting. The DoD has made clear that information is central to the American way of war by designating "gaining information advantage" as a core mission, as well as investing in information systems like the Joint All Domain Command and Control system.[6] The turn to information and its role in systems is globally recognized by ally and adversary alike. China's People's Liberation Army (PLA) has put information technology at the center of its military modernization efforts, with the goal of winning "informatized" wars. The PLA's long-range vision for twenty-first-century warfare goes even further, focusing on "intelligentization," or the application of artificial intelligence to war.[7] Russian military theories espouse similar ideas.[8]

Despite these information-centric visions of future warfare, perfect information games relegate the competition for information to an insignificant sideshow. The way perfect information games misrepresent war and undervalue the competition for information within it exacerbate a host of analytic, conceptual, and strategic pathologies beyond wargaming itself, but the centrality of the problem is often not confronted by the wargaming community.

We are at the cusp of integrating information and information technologies in support of wargaming but have yet to fully embrace them within games themselves. This book is a step in the right direction, but much work remains and the stakes are high. If, for example, the outcome of a DoD wargame hinges on the ability of US forces to suppress enemy air defenses, analysts will likely dive deeper into this mission using more detailed models, threat data, and even live fire exercises with real weapon systems. By systematically misrepresenting the difficulties of collecting, processing, and communicating information, perfect information wargames tend to bias analysis away from these difficult but critical problems, affording them less time and energy than they deserve.[9]

Flawed or incomplete representations of information in wargaming can also bolster bad concepts based on flimsy or invalid assumptions. For instance, an October 2020 wargame designed to test the DoD's Joint Warfighting Concept—its signature operational concept for fighting China or Russia—resulted in a "miserable failure," according to the vice chairman of the Joint Chiefs of Staff, General John Hyten.[10] According to Hyten, the primary problem was that the US "Blue Team" planners tried to gain "information dominance," based on the heroic assumption that US forces could dominate an adversary like the PLA that uses information technologies at least as sophisticated as the DoD's. The fact that this wargame eventually uncovered this dangerous flaw does not outweigh the fact that countless wargames conducted and analyzed over previous years to inform and shape this concept all failed to convey how unrealistic the idea of information dominance would be in a future conflict with China.

Perfect-information games can also create fallacious perceptions for strategists and policymakers. These games exacerbate the misperception that the United States' intelligence, surveillance, and reconnaissance give it a "God's-eye view" of the world. Real-world experience in Iraq, Syria, Afghanistan, and Ukraine contradicts this belief. But unrealistic assumptions abound. These assumptions are especially pernicious in the context of competition or conflict with China and Russia, given these states' abilities to exploit, disrupt, or otherwise interfere with American information systems. These problems are compounded by technology hype, wherein narratives invoke exceptional expectations about future capabilities and operating concepts with little real-world empirical evidence.[11]

IMPERFECT SOLUTIONS

At this point, you might reasonably ask why the defense and wargaming communities often use an inherently flawed method. The most obvious answers are time and its close companion, cost. There are game designs that limit information. The gold standard—at least through the halcyon gaze of nostalgia—is probably the verisimilitude achieved by Naval War College games during the interwar period.[12] According to one of the most respected analysts in modern wargaming, Peter Perla, these games separated players into different rooms to simulate the separation of ships within a broader fleet, and used all manner of physical materials from sheets to large clouds of cotton to simulate the fog of war and smokescreens.[13] Few organizations today—the Naval War College remaining one example—have the time, resources, and staff to conduct games like this.

Simpler double-blind games can be more achievable but still costly. In these games, like our game described below, controllers manage at least three game boards: one that represents the "true" state of the world, one that represents what one player team thinks the world looks like, and another that represents what the opposing team of players thinks the world looks like. This approach requires a small army of personnel to accurately populate, track, and adjudicate three different game boards. For large, complex wargames representing joint military operations—or worse, coalition warfare—this can be a very difficult endeavor far beyond the budgets and capabilities of most wargaming sponsors and practitioners.

The more simplistic block games that inspired Chris allow for imperfect information (e.g., instead of knowing exactly what forces are where on a map, players only know that some forces are in a given location). This method is popular in commercial games. It can be quite effective, adding just enough uncertainty to make players think about what they do and do not know. This method can be difficult to execute in large professional games, however. It requires ensuring that players do not sneak glances at their opponents' forces and managing all the blocks at play can also be overwhelming.

Another common method used in defense wargames is perfect information but imperfect targeting. In this method, both sides can see almost all the forces in play but they cannot strike enemy forces without being able to describe how they would close the "kill chain"—the concatenation of sensors, data networks, platforms, weapons, and processes—needed to target these forces. This method is useful for games designed to scope novel problems, but less so for deeper examination. Perfect information still creates all the pathologies listed above, with imperfect targeting acting as a fig leaf to cover these shortcomings. Kill chains involve more than targeting; the steps required to "find, fix, track, target, engage, assess" are all subject to imperfect information in real life.[14] Giving players the find, fix, track, and assess parts of the chain for free underestimates the difficulty of these functions and the potential ways adversaries could foil them.

At a deeper level, these games oversimplify the challenges of discerning the adversary's center of gravity and allocating scarce resources across space, time, and mission priorities—to say nothing of organizational or domain boundaries—in contested and degraded information environments. China and Russia have spent decades honing their capabilities and concepts for undermining US information in peace and attacking US information systems in war. Chinese military authors write extensively about "three warfares" (legal, public opinion, and psychological) as well as "systems destruction" warfare; Russian authors are no less prolific in writing about "information confrontation" and "perception manipulation/reflexive control."[15] Even if their operations are only partly successful (as indicated by Russian performance in Ukraine in 2022), they stand to make future information environments into a chaotic mess of conflicting information, deception, exploitation, and cognitive overload.

OUR APPROACH

The solution we devised with the Center for New American Security's Gaming Lab was not intended for wargames focused on information—or, for that matter, cyber. We initially planned to run a wargame to examine airpower in a conflict with China. When the COVID-19 pandemic halted in-person games, we pivoted to remote options.

We discovered that remote gaming allowed us to readily control players' information environments. Doing so created new attack surfaces for information operations by opposing teams that our in-person games had lacked. For example, cyber and electronic warfare attacks would not simply happen on the game map and then disappear. Through remote play, these kinds of attacks altered what the players perceived, as well as their ability to communicate or share information. If a team attacked a communication link in the game, for example, its opponents could not communicate in real life. The resulting design applied cyber, space, electronic warfare, and information to conventional wargaming, rather than a game focused on the information environment itself. According to one experienced Air Force player, it was "the most realistic representation of operational command" he had seen in a wargame. Since the first game in early 2020, we have conducted at least ten more games using a similar design.

The Scenario

The objective of the initial game was to explore alternative concepts for airpower in competition and conflict with China (part of a broader project to develop "A New American Way of War").[16] The wargame was set in the 2030s, which was far enough in the future that conceptual change is feasible, but not so far in the future that uncertain technological and political change make forecasting too difficult. The scenario imagined a China-Taiwan crisis stemming from the Chinese Communist Party becoming impatient with the pace of unification.[17] Gameplay and analysis were intended to be at the operational level of war (similarly, see chapter 6).

Structure and Players

We used Zoom videoconferencing software as the primary virtual interface. Additionally, players used Slack to chat and share files, as well as Microsoft

PowerPoint and Google Docs to collaborate on shared plans of action.[18] The game was broken into turns of roughly one to two hours' length in real time. Each turn represented three days of game time. This game, and subsequent iterations, took place over two days of real time.

The infrastructure of the game was entirely digital. We started out by using PowerPoint files of the various maps, counters, and trackers that we would ordinarily print as hardcopies for in-person games. We initially did this for the sake of expedience and, as we have iterated our infrastructure over subsequent games, we have modified the format to be more digital-by-design using Google Docs. It is likely that the user experience of our players could be improved further, but the format worked well enough.

We treated the various software communication and collaboration tools used to play this game as if they were the information and data networks used by military forces in the real world. For instance, we pretended that Zoom equated to high-bandwidth satellite communications for video teleconferencing (not a far stretch). File sharing through Google Docs simulated other military data systems used to provide a common operating picture. Text chat on Slack mimicked low-bandwidth, long-range communications.

Leveraging these data and information networks, our control team could model cyber, electronic warfare, and space attacks in the game. These attacks created "real" effects on the players and teams as their ability to communicate and collaborate was degraded or denied. In later iterations, we added random outages caused by interference, cryptography changes, and other glitches. Several players commented that this approach to degradation was much more realistic than standard games. Even the real-world technical glitches and limitations imposed by the software contributed to the realism of the game, since these same defects exist in real-world command, control, communications, and intelligence systems.

The players of this game and subsequent iterations were a mix of military personnel of various ranks and backgrounds, as well as civilian strategists and analysts from think tanks, academia, and industry. Most of the players had significant experience wargaming similar scenarios, and most of the military players had relevant operational and staff experience. The players were divided into the US / Blue Team and China / Red Team. These teams were further subdivided into the relevant operational and tactical commands involved with this type of operation. For example, the Blue Team had subcomponents for the US Indo-Pacific Command, the Pacific Air Forces / Joint Force Air Component commander, the Pacific Fleet / Joint Force Maritime Component commander, and the US Army Pacific / Joint Force Land Component commander, as well as a forward Joint Task Force commander and forward component commands. Separating the teams into subcommands and placing the groups into separate virtual breakout rooms in Zoom and channels in Slack was critical to represent command, control, and communications pathways and to manage the virtual information environment during the wargame.

Conventional Gameplay

Since analysis of airpower was our original objective, this aspect of the game was more detailed. Player teams allocated aircraft to rotational fixed stations (e.g., combat air patrols), fighter sweeps, or specific strike sorties. Each turn, the air component players submitted an air tasking order that allocated aircraft squadrons and weapons to their missions. The control cell then adjudicated interactions between Red and Blue air forces and actions.

Given the complexity of air interactions (e.g., early warning and target acquisition from several domains, as well as possible fires from subsurface, surface, or enemy aircraft), adjudication used an attrition model that considered mission, aircraft location, geometry of the engagement, random chance, and acceptable risk as defined by the players. Similarly, naval engagements were adjudicated by combining stochastic dice rolls with open-source rulesets like *Harpoon* to determine how many missiles must be fired for the odds to favor hitting a given ship.[19] The videoconferencing and shared data files we used for remote wargaming allowed subject matter experts to adjudicate outcomes at the division/squadron levels for air, brigade level for ground, and ship or task group level for maritime operations.

Information Actions, Effects, and Adjudication

The above-mentioned forces and rule set would be at home in almost any conventional wargame in the DoD. Given remote play, however, our virtual game environment provided a unique opportunity to simulate the information environments involved with command, control, and communications; intelligence, surveillance, and reconnaissance; and precision targeting. The main channels of information consisted of video conferencing, voice chat, and text messaging, as well as several maps that depicted the location of military forces.

These channels could be attacked and the information they conveyed could be degraded or destroyed or include deception. Players chose their attacks from a menu including cyberattacks, electronic warfare, and counterspace attacks. Players at each echelon were provided with a limited number of attack options corresponding to their echelon's span of control; e.g., the Indo-Pacific Command would be able to leverage strategic or national level assets, while the forward component commands' actions were more tactical. The control cell adjudicated the outcomes of these attacks stochastically—mostly with dice—and then applied effects to the opposing team's communications or operational picture.

For example, attacking communications satellites via jamming, laser dazzling, cyberattacks, or kinetic strikes (with possible debris) had a chance of taking down video but virtually no chance of affecting voice or text. In addition to satellite communications, the game also included undersea cables. Players could attempt to physically cut these cables or disrupt them through cyberattacks. As it happened, attacks on undersea cables briefly isolated the US Indo-Pacific Command, which cut off the military commander from their subordinate units and, because communications from Washington ran through Honolulu, also cut off the president from the front lines. More often, Indo-Pacific Command's satellite and undersea-cable high-bandwidth communications would be degraded, meaning that these players' videoconferencing and operating picture would be lost but text and voice would remain. This created an elaborate game of "telephone," in which forward commanders would try to explain what was happening to the Indo-Pacific commander, who would then try to explain it to the president—all over voice or short text messages. As you might imagine, this method was inefficient and prone to miscommunication.

Similarly, this game role-played various component commands, including the Joint Forces Air component commander-forward in Yokota and the Joint Forces air component commander-rear in Honolulu, as well as the commander, Seventh Fleet, in Yokosuka and the Joint Forces maritime component commander in Honolulu. These units often lost communications with the Indo-Pacific Command, its domain counterparts, and its subordinate tactical commands. In partial contrast, the Seventh Fleet's commander rarely lost communication with forward forces, since they played as if aboard a command ship; as a result, they occasionally provided forward air forces with tasking orders in lieu of the air component commanders.

In addition to space, cyberspace, electronic warfare, and undersea cable attacks, the game also allowed kinetic attacks on critical network and command infrastructure. For example, the command ship hosting the Seventh Fleet's commander was easy to locate when communicating, and it was vulnerable to attack when within missile range of Chinese forces. In this game, Chinese cruise and hypersonic missiles launched by stealth bombers and nuclear submarines also had devastating effects on command, control, and communications when they destroyed the Indo-Pacific Command's headquarters, killing both the Indo-Pacific and Pacific Fleet commanders.

To manage the information available to players, the control cell maintained three main maps that showed the different versions of the operational picture. One map represented the true state of the world, seen by control; one represented Blue's picture; and one represented Red's picture. Given imperfect information about the enemy (and reality), these pictures were not the same. Moreover, the maps could be further divided based on subcommands or domain components, representing imperfect information about friendly forces (e.g., the Blue maritime component could be shown a different picture than the Blue air component if they lost their ability to communicate and thus sync their separate operating pictures into a common picture).

To illustrate, the true location of surface ships was displayed on the maps for both teams, unless the players operated them using emissions control to evade detection (i.e., hiding from the enemy by running silent with restricted acoustic and electromagnetic radiation). Adjudicators in the control cell could see where ships using emissions control and submarines were located, but the opposing team could not. Enemy detection of hidden forces was based on their actions, the proximity of sensors (including aircraft and space-based intelligence, surveillance, and reconnaissance), and cyber actions like exfiltrating information from the opposing team. Players could also use decoy ships for deception, in effect cluttering their opponent's picture to complicate or confuse its decision-making.

The upside of using these separate maps was that they allowed for discrete control over what each team could see about what was happening in the game. For example, if adjudicators determined that Chinese cyberattacks had badly damaged the intelligence and communication networks available to the US Indo-Pacific Command, then that subteam would have a particularly flawed picture, even while some subordinates units had more accurate information about the conflict. Clashing operational pictures and degraded communications frequently led to breakdowns in command and control, as players on different subteams tried to reconcile their commander's intent with the tactical and operational situation as they saw it.

The downside of this method is that it was labor and time intensive for the adjudicators. Digital tools for remote play reduced the burden relative to playing such a large game with the wooden blocks described above, but maintaining three or more separate maps was still demanding. Moreover, while the imperfect information in this game was more realistic, it also made more outcomes opaque to the players. With face-to-face adjudication in a perfect-information wargame, the players can more readily interrogate their opponents' actions and the adjudicators' decisions about outcomes. In contrast, in our game, while players could question and potentially change an adjudication decision after the fact, this was the exception rather than the rule, which was also realistic. We do not always know why things happened in real life, nor can we readily change what has happened. However, as suggested in chapter 2, the realism gained through opacity changes aspects of players' engagement.

Overall, managing pictures, video, voice, and text allowed us to produce a nuanced series of effects in the game. When combined with the remote physical locations of players in the roles of different commanders, managing these channels produced a complex information environment. These changes significantly altered the character of conflict.

KEY FINDINGS

The most important finding of this game and subsequent iterations is that, if you aim to simulate modern warfare in a game, you must simulate information environments and the systems that work within them. These environments include dependent functions like command and control; intelligence, surveillance, and reconnaissance; and targeting. In analytical wargaming for and about the DoD, it is not enough to give players "cyber bombs" and then proceed to play an otherwise-conventional wargame with perfect information. This approach, which we argue is common in today's defense wargaming practice, undersells and distorts the value of information operations—including cyber.

When cyber is simply "sprinkled on" otherwise conventional games, players and military planners tend to treat cyberattacks and electronic warfare as "soft kill" substitutes for kinetic actions. They are less inclined to use these attacks to penetrate and disrupt their opponent's thinking and perception of the world. These two types of attacks are related but distinct, as are intelligence, surveillance, and reconnaissance or targeting operations. They should not be used in the same way. However, when wargames treat information and influence operations as if they have no meaningful effect, it is rational for players to use their cyber "bombs" (or electronic warfare and space "bombs") just as they would use regular bombs or bullets for hard kills. This is problematic.

Building information challenges of command, control, and communications into wargaming can be difficult, but it is the only way to realistically represent the character of modern war and the challenges that future policymakers and commanders will likely face. At a bare minimum, these representations should address the fundamental challenges of understanding the disposition of hostile and friendly forces, executing kill chains, and exercising command and control. Building a wargame without representing information, targeting processes, or command and control is as unrealistic as building a wargame without fighter aircraft or submarines or air defenses. We would rightly ridicule such an approach, and yet many in the defense and wargaming communities tolerate this practice. It is a blind spot that, if left unaddressed, will continue to feed dangerously inaccurate thinking.

Our wargame saw the US players *attempt* to execute some manner of multi-domain operations through space, cyberspace, the electromagnetic spectrum, air, sea, undersea, and land to create synergistic effects on a target. While this idea sounds great, its execution in our game was a mess. As communications degraded and shared pictures diverged, each service- or domain-specific command started operating independently, not as a piece of a broader joint or all-domain campaign. These commands naturally gravitated toward familiar operations. Navy forces executed a fleet battle. Air forces executed an air battle. These two were almost entirely disconnected. This is not to say that all forms of joint or multidomain integration are impossible or foolish. Nevertheless, the information and command-and-control assumptions behind these concepts need critical examination to see how they function under the fog and friction of war with a near peer adversary.

Finally, remote wargames offer remarkable promise for innovation. The game described above only scratched the surface of what is possible. Subsequent iterations improved on the basic design; however, virtual game platforms using purpose-built software stand to offer designers and adjudicators even greater control over information while also providing players with more realistic representations of imperfect-information modern warfare. Similar infrastructure may also provide for better data capture and analysis. These opportunities, coupled with remote gaming's potential for collaboration around the globe without expensive travel, could usher in a new era of wargaming.

NOTES

1. Philip Sabin, *Simulating War: Studying Conflict through Simulation Games* (New York: Bloomsbury Academic, 2012), Kindle edition, location 2586.
2. Andrew Parks, "Napoléon: The Waterloo Campaign, 1815," BoardGameGeek, no date, https://boardgamegeek.com/boardgame/1662/napoleon-waterloo-campaign-1815.
3. John F. Sullivan, "Interpreting Sun Tzu: The Art of Failure?" *Strategy Bridge*, July 13, 2021, https://thestrategybridge.org/the-bridge/2021/7/13/interpreting-sun-tzu-the-art-of-failure.
4. Carl von Clausewitz, *On War*, edited by Michael Howard and Peter Paret, vol. 117 (Princeton, NJ: Princeton University Press, 1976), 88.
5. For more on Lanchester Square equations, see Alan Washburn, "Lanchester Systems" (Monterey, CA: Naval Postgraduate School, 2000).
6. John R. Hoehn, "Joint All Domain Command and Control (Jadc2)," Congressional Research Service, 2020; Alexus G. Grynkewich, "Introducing Information as a Joint Function," US Air Force, 2018.
7. Edmund J. Burke, Kristen Gunness, Cortez A. Cooper III, and Mark Cozad, "People's Liberation Army Operational Concepts," Research Report RR-a394-1 (Santa Monica, CA: RAND Corporation, 2020), 21.
8. Timothy L. Thomas, "Russian Military Thought: Concepts and Elements," MITRE Corporation, 2019, 8–12.

9. This is likely difficult to prove without a detailed assessment of DoD and Service analytic priorities, but it comports with the allocation of spending in the defense budget in which spending on platforms and kinetic weapons systems dwarfs spending on information systems and weapons. See Andrew Eversden, "What the Budget Reveals—and Leaves Unclear—About the Cost of JADC2," C4ISRNet, June 17, 2021, www.c4isrnet.com/c2-comms/2021/06/15/part-1-what-the-budget-reveals-and-leaves-unclear-about-the-cost-of-jadc2/.

10. Tara Copp, "'It Failed Miserably': After Wargaming Loss, Joint Chiefs Are Overhauling How the US Military Will Fight," *Defense One*, July 28, 2021, www.defenseone.com/policy/2021/07/it-failed-miserably-after-wargaming-loss-joint-chiefs-are-overhauling-how-us-military-will-fight/184050/.

11. On this concept, see Frank L. Smith III, "Quantum Technology Hype and National Security," *Security Dialogue* 51, no. 5 (2020).

12. John M Lillard, *Playing War: Wargaming and US Navy Preparations for World War II* (Lincoln: University of Nebraska Press, 2016).

13. Peter P. Perla and Ed McGrady, "Why Wargaming Works," *Naval War College Review* 64, no. 3 (2011).

14. See John A. Tirpak, "Find, Fix, Track, Target, Engage, Assess," *Air Force Magazine*, August 20, 2021, www.airforcemag.com/article/0700find/.

15. For the Three Warfares, see Elsa Kania, "The PLA's Latest Strategic Thinking on the Three Warfares," *China Brief* 16, no. 13 (August 22, 2016), https://jamestown.org/program/the-plas-latest-strategicthinking-on-the-three-warfares/. For Russia, see Keir Giles, "Handbook of Russian Information Warfare," Fellowship Monograph 9, NATO Defense College, November 2016, 3–5, www.ndc.nato.int/news/news.php?icode=995; and Timothy L. Thomas, "Russian Military Thought: Concepts and Elements," MITRE Corporation, August 2019, 9.1–9.29, www.mitre.org/sites/default/files/publications/pr-19-1004-russian-military-thought-concepts-elements.pdf.

16. Christopher M. Dougherty, "Why America Needs a New Way of War," Center for a New American Security, 2019, www.cnas.org/research/defense/a-new-american-way-of-war.

17. For more on the 1992 Consensus, see Derek Grossman, "Is the '1992 Consensus' Fading Away in the Taiwan Strait?" RAND Corporation, June 3, 2020, www.rand.org/blog/2020/06/is-the-1992-consensus-fading-away-in-the-taiwan-strait.html.

18. All these platforms are often readily available and very familiar to most players; however, it is a good wargaming practice to provide a step-through guide that allows players to familiarize themselves with the technologies before gameplay.

19. "Harpoon V," in "The Admiralty Trilogy," no date, www.admiraltytrilogy.com/harpoon.php.

6

ADDING TIME TO THE CYBER KILL CHAIN: THE "MERLIN" TOOL FOR WARGAMING

PAUL SCHMITT, CATHERINE LEA, JEREMY SEPINSKY, JUSTIN PEACHEY, AND STEVEN KARPPI

Success in modern warfare requires integrated military operations in, across, and through all domains. The cyber domain and information environments are interdependent with the purely physical domains of land, sea, air, and space. However, as in the real world, the uncertainties that surround cyber actions and effects can make them difficult to integrate into multidomain wargames. Many of the tools used to represent cyber actions through a series of "moves" during a wargame are still in their infancy; so too are attempts to adjudicate their effects and operational implications. Incorporating cyber actions and effects in wargames with concurrent moves in physical domains is even more challenging, as is determining probabilities for likely outcomes. More often than not, the resources and timelines needed to prepare and execute such actions in the real world are greatly underappreciated.

This chapter examines one innovative solution to these challenges for cyber wargaming, called Merlin. Developed by the Center for Naval Analyses (CNA), Merlin is a tool that models the resources needed to develop and implement the cyber kill chain over time.[1] In the legend of King Arthur—and in T. H. White's classic book *The Once and Future King*—Merlin "was born at the wrong end of time." The mythological wizard experiences life in the reverse order from the other characters in the story. The past is his future, and the future his past. Similarly, the Merlin cyber wargaming module challenges cyber operators to plan backward into the past.

As its namesake suggests, this tool introduces some wizardry to shape the future from the past. First, real cyber exploits take time to build. They are not simply "available on tap." Therefore, based on real world cyber operations that the CNA has reconstructed, Merlin identifies the amount of effort needed to "backcast," namely, retroactively posit the prospect that any given cyber exploit would be operational for players to use during gameplay. Second, real cyber exploits can fizzle out or be discovered. So, Merlin also identifies the probability that a given exploit will work as intended.

Said another way, Merlin is an ex post facto—or retroactive—tool for generating a synthetic history. This means that, when used in a wargame, it can create a cyber development history after the facts have arisen in gameplay. Players ask, "based on what we need today," in game time, "what would we have previously done to provide what we now need, exactly when we needed it?" Merlin adds some realistic constraints to cyber actions during wargames while still allowing players flexibility and creativity in their fictional campaign planning. Data from the Merlin module can also be analyzed after the wargame to examine critical planning elements for cyberspace operations. Among others, these elements include the potential disconnect between cyber planners and conventional operators; the amount of time and effort required to devise and deliver a cyber effect at the time and place of a player's choosing; and the type of offensive and defensive effects that may be useful in a given theater of operations.

Merlin is intended for operational or campaign-level wargames. The operational level of warfare links the tactical employment of forces up to strategic objectives.[2] Tactics involve individuals or small units, including the specific steps they take to attack and defend specific targets. Tactical actions generally occur on time scales of minutes, hours, or days. At the operational level, campaigns are developed and executed across a theater of operations. They involve larger units of force, such as fleets and armies. The timelines for these operations are more protracted, lasting days, weeks, or months. The strategic level further aggregates these campaigns across longer periods of time (e.g., weeks, months, or years), larger spaces (e.g., regions or the world), and all levers of national power (e.g., diplomatic, informational, military, and economic).

These distinctions provide an important mental model for decision-making about national security in general, and cyber capabilities in particular. The insights developed by wargaming an operational environment (represented by the game board), force structure and capabilities (game pieces), and net assessment of outcomes from their interactions (adjudication) are what enable analytical wargames to inform real investments to improve operational capability and, ideally, improve integrated campaign planning.[3]

The Merlin tool can be abstracted to support cyber wargaming at different levels of security classification—including the unclassified applications discussed in this chapter. In this context, Merlin can support not only research or analytical wargames but also educational wargames for students in professional military education and other settings who are learning about cyber operations and planning. Emphasis is placed here on Merlin as an analytical tool.

BACKGROUND ON MERLIN

The Center for Naval Analyses developed the Merlin module for the Air Force Research Laboratory Strategic Planning and Transformation Division. The module was designed in consultation with subject matter experts in cyberspace, including those with backgrounds in offensive and defensive cyber operations, as well as the design and development of cyber tradecraft.

This chapter describes Merlin as a tool to support two-sided, force-on-force wargames with campaign level effects. The module itself is not a standalone wargame (although it has recently been expanded to create a freestanding game called "Merlin in a Box"). As described here, Merlin is a supplementary tool that provides a vocabulary, taxonomy, and venue that brings cyberspace operators and traditional warfare operators together as players in the same game. Using this tool, these players can design, deploy, and integrate nonkinetic cyber tradecraft with kinetic operations to achieve military objectives in the game.

CNA also developed a Merlin adjudication rubric to aid the analysis of players' cyber tradecraft. This rubric helps researchers and players alike evaluate several reoccurring questions about cyber operations. For example, what kinds of cyberattacks are effective, and why? What are the development timelines for offensive and defensive cyber operations? Similarly, what are the labor requirements for these operations? And what are the barriers to effective offensive and defensive cyber operations? The adjudication rubric provides an opportunity for a robust discussion of both the opportunities and limitations of integrating cyber operations with operations in other domains.

BETTER TOOLS TO EVALUATE CYBER ACTIONS

Many wargames use simple representations of cyber capabilities, such as action cards, preset menus, and game points. These devices can help identify players' priorities. They can also help to suppress the appetite of uninformed players to launch unrealistic cyberattacks on untenable timelines.

However, many military planners and decision-makers need a more sophisticated appreciation for the time it may take to (1) collect and analyze the intelligence needed to understand a potential target, including reconnaissance to identify cyber target entry points; (2) gain access to that target through cyberspace; and (3) develop and employ the weaponized cyber tradecraft that could achieve the desired effect on the target.[4] They also need to appreciate the risks that these cyber actions could be discovered and crippled by their adversary (i.e., burned).

Merlin helps address these nuanced but nevertheless critical aspects of cyber competition and conflict. It incorporates five distinct aspects of cyber tradecraft development and execution, which are often overlooked and insufficiently explored in wargames not dedicated to cyberspace. First is operational flexibility for the warfare commander. Merlin lets players in the role of warfare commanders choose the effects they want as they need them, instead of requiring them to plan in advance. This design sacrifices some realism. However, it facilitates dialogue among players about what might be useful (instead of only the effects that subject matter experts are prone to consider), thereby encouraging creativity that supports education as well as new ideas that may be useful in further exploration of emerging warfare-related concepts.

Second are realistic constraints on tradecraft development. As players request cyber effects in the wargame, Merlin records and tracks the historical personnel requirements that these move requests would generate. Facilitators in the Game Control Cell compare these requirements with empirically supported personnel caps from a historical database that corresponds to the development timeline for the requested cyber effects. The personnel requirements are then projected into the past of the game timeline. As a result, players using the Merlin module will determine the resources and timelines (potentially years long) needed to prepare and execute the cyber actions they request. In some cases, this means that the proposed action needs to be scoped down, or the probability of success needs to be lowered, or the date of execution needs to be pushed later into the game.

Third is flexible cyber defensive posturing. Players are given an opportunity to defend various aspects of their own information technology infrastructure. They can allocate defensive resources to hardening and resiliency, which decreases the likelihood of a successful attack by an adversary. Alternatively, they can invest in sensors and forensics, which increase the probability that their adversary's offensive tradecraft will be discovered and burned. Players can choose which targets are defended and the levels of defense; facilitators can choose to set the amount of cyber defense available based on the country each player is roleplaying. This encourages an understanding of both offense and defense, improving analytical outcomes for cross-domain warfare. It also provides educational benefits.

Fourth is the success or failure of offensive moves. Each piece of offensive tradecraft developed is given a base probability of success—$P_{success}$—depending on the cyber target and the accesses used to attack it. This can be modified by the complexity of the desired effect, as well as the defensive systems emplaced by the adversary. Success or failure, based on this probability, is determined only when a player or team attempts to execute the desired effect. This stochasticity addresses the volatility and range of uncertainty associated with cyber effects while also providing some basis for player planning and decision-making on risk assessment.

And fifth is the discovery of offensive moves. Each piece of offensive trade-craft that is executed on an adversary's system can potentially be discovered. Multiple executions of the same tradecraft increase the probability of discovery—P_{burn}. This separates the tradecraft itself from the effect that it can generate.

As the fourth and fifth element indicate, cyber offense warrants special attention. This is in part due to the assumed advantage of "initiative," which is typically afforded to the offense. For example, in naval warfare, Wayne Hughes advocates that commanders "fire effectively first" (even while he acknowledges the importance of "choosing the right moment to attack, neither too soon nor too late").[5] In cyber warfare, the relative advantages of offensive versus defensive cyber operations are hotly debated.[6] In wargames, adjudication of force interactions requires a net assessment of offensive and defensive actions to determine outcomes for effect achieved. Characterizing cyber offense is a reasonable place to start. After all, there has been considerable work on analytical tools that describe the kinetic munitions used in the physical domains (i.e., missiles and bombs), as evidenced by the *Joint Munitions Effectiveness Manual* (*JMEM*). But little work has been done to characterize planning for cyber munitions into a "cyber *JMEM*."[7]

BACKCASTING TO ACCOUNT FOR RESOURCES

Using Merlin, cyber effects are created via "backcasting."[8] This construct assumes that cyber planners in the wargame have perfect foreknowledge of what military commanders will desire during a future conflict. Those planners are also assumed to have created the correct cyber tool, which will be available at the moment of need. While these are heroic assumptions, they are also useful. They allow players to develop and deploy cyber tradecraft that are creative and appropriate for the task at hand. An analytical benefit of this methodology is that players may identify useful tradecraft that had not been previously considered, which may in turn inform future decision-making about warfighting concepts, investments, and resource allocation in the real world.

Backcasting allows cyber tradecraft to be relevant in gameplay in a realistic fashion. It forces players in the role of cyber operators to continually pay attention to the wargame and actively participate in the development of useful tradecraft. When the wargame is over, all players will have a record of cyber tradecraft used during the wargame, as well as an understanding of their likely effectiveness. Players can describe their cyber requirements based on the effects they desire, instead of forcing them to decide in advance what cyber tools will be needed.

IMPROVED ADJUDICATION

Other wargame constructs create a set of cyber tradecraft at the start of the wargame. Typically, these games provide either a menu of actions and effects from which players can choose, giving them a fixed number of cyber "bullets" or a basket of cyber "points" to pay for various actions. These resources must be judiciously allocated, and most are assumed to be ready for use at the start of the game. Players must define their cyber targets ahead of time and then expend their cyber "readiness" for actions in a given move, usually in a "use or lose" concept of employment. Alas, many of these games do not account for potential increases in readiness over time (as can happen in real life), nor the potential for players to expend their readiness to achieve a desired cyber effect more dynamically.

In partial contrast, during a wargame using Merlin, players fill out a cyber effects card—including information operations that are disseminated via cyberspace—and provide it to the Merlin adjudicator in the Game Control Cell. The adjudicator uses the Merlin Spell Book to capture and build out a record of "casting spells" (playing on the magic theme), namely, cyber moves.[9]

To create the tradecraft and provide the desired effect, players define an Access Approach Class (AAC) and a Cyber Target Class (CTC). We describe these below. Each combination of AAC and CTC generates a set of personnel resource requirements. These requirements are determined in part by dice rolls, thereby providing some randomization in the amount of human capital and time that are needed to develop the cyber tradecraft in question. The Spell Book "backcasts" these personnel requirements over time and across a number of different personnel talent management types. These types might include, for example, joint intelligence operations centers. There is a limit to the maximum number of cyber exploits that each personnel type can create simultaneously. The key data thread is the labor constraint to meet the requirement.

As players develop multiple cyber tools for employment, the personnel requirements at a given point in the past may exceed the personnel resource caps. In this case, the development of cyber capabilities will need to be shifted, either into the past or the future, in such a way that no resource caps are exceeded. The Merlin adjudicator assists in shifting exploit development. It is important for these shifting timelines to support analytical wargaming on operational planning, as well as resource allocation to support force planning; they are also important for professional education, helping players to understand whether their move could plausibly be accomplished in a resource-constrained environment.

It may take multiple turns of the wargame before exploits shifted into the future will be available. This realistically constrains the high demand for cyber actions sought to satisfy a voracious player's appetite to "cyber everything." When a piece of tradecraft has been developed, the Merlin adjudicator will fill in the "Executable After Turn," "P_{burn}," and "$P_{success}$" boxes on the cyber effect card and then return this card to the player team that developed it.

SETUP AND PLAY

In wargames that use Merlin, each of the player teams typically has a leader who coordinates the team's actions and synchronizes their intended effects. The teams may also have players who are subject matter experts for regions or domains of the game's "battlefield." One or more of these players form the Cyber Cell, focused on the effects to be achieved in or through the cyber domain. The controllers are the referees who control the gameplay.[10] These referees work with the adjudicators, who judge the outcomes of a team's planned moves against the opposing team's planned moves. Generally, the two teams are balanced in size and expertise to stimulate the best interaction and most challenging competition.

Before executing a wargame with this tool, controllers must set up a few game pieces. AACs are methods for gaining access to a cyber target in order to implant a piece of tradecraft. CTCs are the systems being targeted by cyber tradecraft. Controllers must define the set of AACs and CTCs available for gameplay. They must also initialize the Merlin Spell Book with appropriate data. These data can be informed by deductive assumptions or empirical case studies, tailored to the scenario, players, security classification, and game objectives (e.g., what the game sponsors define as the research question or what participants want to learn).

Access Approach Class Cards and Cyber Target Class Cards

An illustrative AAC card is shown in figure 6.1. Similarly, see figure 6.2 for a CTC card. Players develop and submit their move by placing two of these cards on top of each other, with the AAC card on top. This shows the desired cyber effect against the intended target, the tradecraft tool to be used, the method for access, and other relevant information in a quick glance.

After a player arranges and submits the paired AAC and CTC cards, a Merlin controller fills in the required Spell Book probabilities. Next, they hand off

 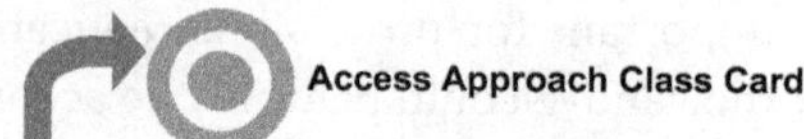 **Access Approach Class Card**

> Depending on access sought (e.g., remote access), populate this field with personnel requirements for development, with timeline at each step to be determined by dice rolls.

Tradecraft Name:		
Tradecraft available as of:		
Desired Effect:		

Turn	$P_{success}$		P_{burn}	
	Base	Roll	Base	Roll
	80%		20%	
			+20%	
			+20%	
			+20%	
			+20%	

FIGURE 6.1. Sample Approach Access Class (AAC) Card.

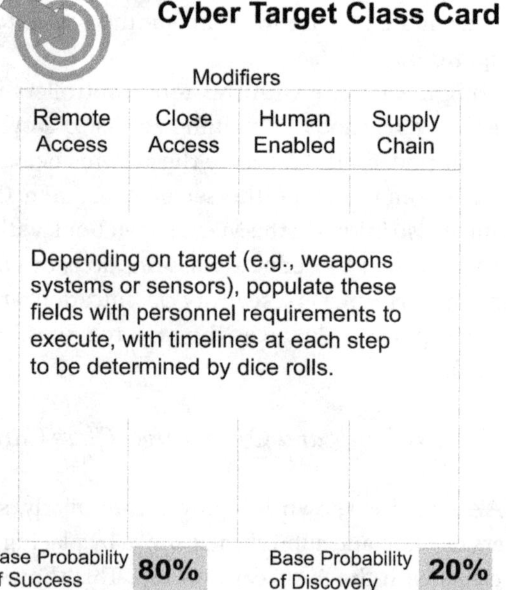

FIGURE 6.2. Sample Cyber Target Class (CTC) Card. This card illustrates how different kinds of cyber access modify the dice rolls that determine the probably of success and discovery.

the stapled pair of cards to the controller team for integration with other proposed player actions, which may be concurrent in other domains. This handoff indicates that the cyber action described is Merlin approved and ready for execution.

Cyber Offense: Merlin Adjudication

There are five steps for offensive cyber adjudication to create and execute a new piece of tradecraft. In the next subsections, we walk through an illustrative example.

STEP 1: CYBER TARGET SELECTION

Once the Cyber Cell players determine a desired cyber effect (or are asked for such an effect by their team), they decide what target(s) they need to attack to achieve that effect. They develop a plausible narrative that describes the desired impact on the adversary, what kinds of tradecraft are being used, how the tradecraft will lead to the desired cyber effects, and what the observable outcomes would be if successful. Facilitators may help players articulate their intent. Descriptions of the tradecraft should be as specific as required by the Merlin controllers and, of course, will be limited by the classification of the wargame environment.

For example, imagine that the desired cyber effect is a traffic jam in downtown Chicago. The narrative might be as follows: Let us call the tradecraft name "Traffic Disruption." Cyber operators will gain real-time remote access to the integrated network of traffic monitoring cameras at the Illinois Department of Transportation. This will allow them access and visibility of real-time traffic data, as well as the ability to control the timing and sequencing of traffic lights. Cyber operators will extend and delay the red/green cycles of lights in the desired region to create gridlock across major intersections. Less than 10 minutes of modifications should generate severe traffic congestion. Traffic controllers at the Transportation Department will likely notice the congestion and will be working actively to mitigate the impact.

Once the target narrative has been created, Cyber Cell players work with the Merlin controllers to determine which CTC most closely fits the target affected by their narrative. The CTCs should be a well-researched taxonomy of the kinds of targets that have been affected via cyberspace in real life (or could be). A sample CTC taxonomy might include weapons systems, national or military sensors, national or military command and control / computers and communications infrastructure, critical military or civilian infrastructure and key resources, cyber actors' infrastructure, Internet information technology, or information/media sources and services (see figure 6.2).

Each target narrative should fit a single CTC. If more than one CTC is appropriate, Cyber Cell players can choose the CTC they prefer. If multiple CTCs must be affected to achieve the desired cyber effect, multiple tradecrafts will likely be required. If a cyber effect is particularly valuable or critical, Cyber Cell players are free to develop multiple independent tradecrafts (subject to available resources) to provide redundancy for execution on the same system via the same access approach. This could provide a higher probability of the effect being successfully achieved. If more than one target must be affected to create the desired effect, each independent class of target is considered a separate cyber effect.[11] CTC cards are provided to the Cyber Cell players to summarize the relationship between the CTCs and the AACs. The chosen CTC may modify the timeline, difficulty, or discoverability of a given tradecraft.

STEP 2: SELECT ACCESS APPROACH CLASS

Building on the narrative created in step 1, Cyber Cell players next need to determine how they supposedly gained access to the cyber target. Cyber Cell players are encouraged to generate a plausible access narrative.

For example, continuing the earlier narrative about a cyberattack on Chicago's transportation system: Cyber operators began with a broad-spectrum phishing campaign against city and state employees. They used information gained through that effort to spear-phish user accounts with known accesses to the traffic control systems. Once an appropriate account was accessed, offensive cyber teams exploited a backdoor to allow real-time remote access. A taxonomy of AACs includes remote access (from geographic sanctuary), close access (from inside adversary-controlled or neutral territory), human-enabled access (from inside an adversary facility), and supply chain (indirect) access.

STEP 3: ROLL MERLIN DICE

Steps 1 and 2 could both be considered planning exercises. They set up and determine what actions must be taken to create a desired effect. During step 3, resources must be allocated for the selected method of attack against the selected target.

To do so, the dice are rolled. Merlin controllers will input the dice roll into the Spell Book to stochastically determine the time required for necessary personnel to develop and deploy the selected tradecraft to achieve the desired effect. The costs of execution are considered nonrefundable. At this step, the tradecraft will have been developed, even if the costs are higher than expected or desired, if it will not be available when needed, or if the players in the role of warfare commanders never use its effects. Since there is high variability in real-world timelines for developing cyber tradecraft, we want to ensure that

Cyber Cell players commit to a decision before they know the actual development times.

STEP 4: UPDATE THE TIMELINE

Once the Merlin controllers have entered the dice rolls from step 3 into the Spell Book, the timeline is updated. The Spell Book will display the new exploit in the timeline, along with any other exploits already under development. The Merlin controllers will then inform the Cyber Cell players of any conflicts with their tradecraft development.

STEP 5: DETERMINE $P_{SUCCESS}$ AND P_{BURN}

After the timeline has been updated, the cyber tradecraft is ready for use. However, being ready for use does not guarantee success. Each combination of AACs and CTCs has a base probability of success, which is $P_{success}$, displayed in the Merlin Spell Book and on the CTC card. Merlin controllers apply any appropriate modifiers to this number, based on the difficulty of the desired effect or extenuating circumstances to determine the final $P_{success}$. This percentage should be provided to the players in the Cyber Cell before their team executes the cyber action.

The timelines developed in step 4 and the probabilities in step 5 from the Spell Book may seem magical or mysterious to some players. Ideally, the values are derived from empirical data, such that as case studies of real-world events. Here, as elsewhere, some players may fight the scenario (or model), based on their subject matter expertise and experience. This is not all bad. Players' knowledge can be used to update or improve the timelines and probabilities that have been applied. However, players must learn that they cannot just "sprinkle some cyber" onto their campaign, and the development timeline is considerably longer than the tactical planning horizons.

Cyber Defense: Merlin Adjudication

Before the wargame begins, Cyber Cell players assign their cyber defenses. These are described in two categories: (1) hardening/resiliency; and (2) sensors/forensics.

Hardening/resiliency represents the design-level diversity of their team's cyber infrastructure, the redundancy of the available hardware, and any active measures in the system that prevent unauthorized access. Increasing the hardening/resiliency in a given area decreases the likelihood that an adversary's offensive cyber effect will be successful.

Sensors/forensics represents the tools available to cyber defenders to identify what has happened after their system became compromised, to mitigate the

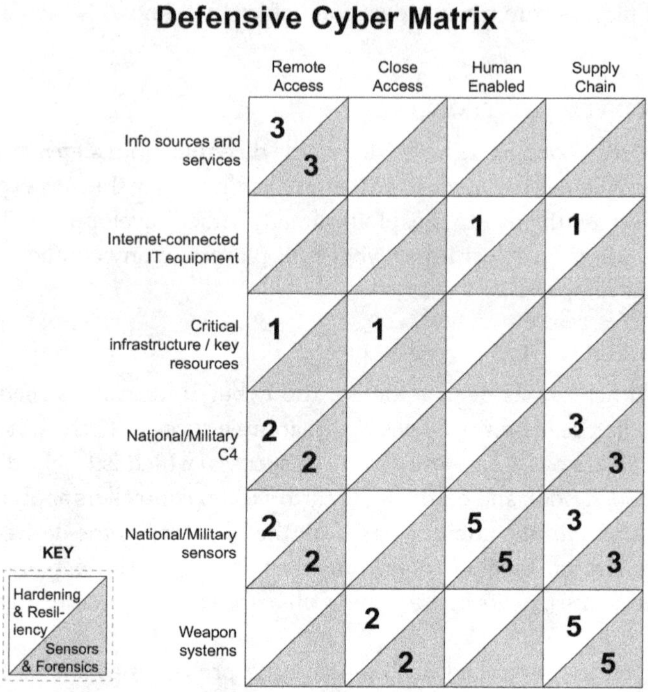

Defensive Cyber Matrix

	Remote Access	Close Access	Human Enabled	Supply Chain
Info sources and services	3 / 3			
Internet-connected IT equipment			1	1
Critical infrastructure / key resources	1	1		
National/Military C4	2 / 2			3 / 3
National/Military sensors	2 / 2		5 / 5	3 / 3
Weapon systems		2 / 2		5 / 5

KEY

Hardening & Resiliency / Sensors & Forensics

FIGURE 6.3. Sample Defensive Cyber Matrix

vulnerabilities, and to patch them. Increasing the sensors/forensics defenses in a given area increases the likelihood that, if a cyberattack against the team was successful, it will at least be discovered and the tradecraft will be burned.

For each combination of CTCs and AACs, each player team usually receives two defense points. Figure 6.3 shows a sample defense matrix. Note that each cell in the matrix is divided diagonally into two blocks. The upper left of each cell holds the number of points assigned to hardening/resiliency. The lower right of each cell holds the number of points assigned to sensors/forensics. Cyber Cell players assign their points to any desired section of the cells. For example, if there are forty grid boxes, each with an upper and a lower half, a total of eighty defense points can be distributed as the players desire throughout their defense matrix. Players may leave cells empty to concentrate defenses elsewhere. If a cell is empty, then the base probabilities of success and discovery are applied.

Similarly, each combination of AAC and CTC cards also has a base probability that the offensive tradecraft will be burned when used, which is P_{burn}. This probability is the likelihood that, when it is utilized for a cyber effect, players on the opposing team will notice the cyber intrusion, mitigate the exploited cyber vulnerabilities, and prevent the exploit in question from functioning again.

At the discretion of the Game Control Cell, cyber effects that are obvious (e.g., attributable, kinetic, or observable) can increase P_{burn} by up to 50 percent. Changes to the base probabilities should be determined by the Merlin controllers based on conversations with players in the Cyber Cell, subject matter expertise, or other inputs. The goal is to reflect, as accurately as possible, the chance of successfully creating the effect or losing the capability. Merlin controllers have the final say.

Finally, cyber defenses must be applied to generate the ultimate probabilities of successful or burned tradecraft. After the above modifiers are applied to the tradecraft $P_{success}$ and P_{burn}, Merlin controllers look at the cyber defenses of the target of an offensive AAC/CTC combination, as illustrated in figure 6.2. For each point the adversary has spent in hardening/resiliency for that combination, controllers reduce $P_{success}$ of the offensive tradecraft by 5 percent. For each point the adversary has spent in sensors/forensics in that combination, controllers increase P_{burn} by 5 percent. When these offensive and defensive elements are taken together, Merlin provides wargamers with a flexible cyber development and adjudication mechanism.

SUMMARY

Wargaming is an important methodology for research and education, as suggested throughout this book. Relatedly, it is useful for informing operational warfighting planners and commanders about the potential strengths and weaknesses of different courses of action, given the available forces on a virtual or physical battlefield during a military campaign. In this way, cyber wargaming can be an important inductive research method for gaining insights into investment and resource allocation decisions that shape the design and architecture of a military force—ideally, to align with the ends, ways, and means of national strategy by prioritizing what capabilities to buy, hold, or further divest.

Merlin advances the state of cyber wargaming by providing a vocabulary, taxonomy, and venue that brings together as players both cyberspace operators and traditional warfare operators to examine how effects integrate across domains. It is primarily described here as an analytical tool for research. However, Merlin is also useful for educating professionals in national security, improving their understanding of the integration of both nonkinetic and kinetic effects. Having knowledgeable practitioners play as subject matter experts in the Cyber Cell adds significant value to the tool's utility in both analytical research and educational games.

This chapter has discussed concepts and processes for conducting net assessments in wargames to determine plausible outcomes of the interactions between cyber-capable adversaries, to include the integration of cyber effects across domains. Even more uniquely, the Merlin cyber module also imposes the realism of time and resources to develop and implement offensive cyber tradecraft, and the potential for this tradecraft to be burned.

Of course, there is always room for improvement. For example, the version of Merlin described in this chapter focuses more on offense than defense. However, there is more to cyberspace than cyberattack options, and more could be included in the model. The same goes for including unforeseen and perhaps unintended consequences of cyber actions. That said, by incorporating critical considerations of time, resources, and tradecraft, the version of Merlin described here significantly improves the tool kit available for studying the operational level of (cyber)war.

NOTES

1. Also, the creators felt that *Merlin* was an appropriate moniker because cyber tools are indistinguishable from magic for many operators. "Any sufficiently advanced technology is indistinguishable from magic," according to Arthur C. Clarke's Third Law, from *Profiles of the Future: An Inquiry into the Limits of the Possible*, rev. ed. (New York: Harper & Row, 1973). Also see Jeremy Sepinsky, Eric Huebel, and Matthew Cupian, "Gaming Cyber in an Operational-level Wargame: Merlin Cyber Wargame Module Rules for Adjudicators," Center for Naval Analyses Research Memorandum DIM-2018-U-018842-FINAL, January 2019; substantial portions of this chapter include excerpts from this report, and all these excerpts are unclassified and have been deemed publicly releasable.
2. Joint Publication 3-0, Joint Operations, January 17, 2017, incorporating change 1 of October 22, 2018.
3. Keith F. Joiner, "How Australia Can Catch Up to US Cyber Resilience by Understanding That Cyber Survivability Test and Evaluation Drives Defense Investment," *Information Security Journal: A Global Perspective* 26, no. 2 (2017): 74–84, doi:10.1 080/19393555.2017.1293198.
4. See https://prevalent.ai/resources/tradecraft-putting-the-security-back-into-cyber security/. The term "tradecraft" here describes practices, skills, and associated tools used in cyber operations acquired through experience in specific techniques to achieve cyber effects. The tradecraft can be used in a mission projects' power and offensively attack an adversary, or to defend one's own cyberspace against an ongoing or imminent cyberspace threat.
5. Wayne P. Hughes, CAPT, USN (Ret.), and Robert P. Girrier, RADM, USN (Ret.), *Fleet Tactics and Naval Operations*, 3rd ed. (Annapolis: Naval Institute Press, 2018), 30–34.
6. Among others, see Rebecca Slayton, "What Is the Cyber Offense-Defense Balance? Conceptions, Causes, and Assessment," *International Security* 41, no. 3 (2017): 72–109, doi:https://doi.org/10.1162/ISEC_a_00267.
7. Mark A. Gallagher and Michael C. Horta, "Cyber Joint Munitions Effectiveness Manual (JMEM)," *American Intelligence Journal* 31, no. 1 (2013): 73–81. This discrepancy is not unique to cyber. See Frank L. Smith III, *American Biodefense: How*

Dangerous Ideas About Biological Weapons Shape National Security (Ithaca, NY: Cornell University Press, 2014).

8. A "cyber effect" is defined herein as any effect generated via the cyberspace domain. The same effect may be created via other domains. However, the Merlin module focuses exclusively on tracking and adjudicating effects generated via the cyber domain.

9. The Merlin Spell Book was originally a Microsoft Excel spreadsheet companion tool for the Merlin cyber module that calculated the required dice for each combination of AAC/CTC cards in a proposed tradecraft move; created the historical personnel timeline based on the player dice rolls; and retained a record of the tradecraft developed. More recently, the Spell Book has been rebuilt as a Python-based tool.

10. Merlin Game Controllers are the members of the "White" or "Control" Cell. The wargame Facilitators are responsible for coaching and assisting players in each team's Cyber cell with Merlin adjudication and the inputs into the *Merlin Spell Book*. Facilitators, Controllers, and Adjudicators are all in the Control Cell; they may be the same or different individuals, depending on game size and complexity.

11. E.g., if a cyber effect requires shutting down a radar system and a traffic system, these would be two separate cyber effects since they fall into two distinct CTCs and the two systems are entirely independent of each other. If, instead, a cyber effect requires corruption of a radar database and an interruption of signal processing in the same radar system, it is plausible to assume that the same accesses and/or similar tools can be employed to generate the result. In the latter case, only one trade craft would be required while, in the former, two sequences must be performed, resulting in the generation of two separate tradecrafts.

7

GAMES WITHIN GAMES: COGNITION, SOCIAL GROUPS, AND CRITICAL INFRASTRUCTURE

RACHAEL SHAFFER

Why do people build cyber wargames? And why do players make the decisions they make during these games? As this book argues, the purpose or objective for a wargame is paramount to everything else that follows, from the selection of players and the scenario to the cyber actions taken and adjudication of their effects. Be they for research or education, the actual and underlying motivations for wargames are not always self-evident. The various reasons why wargames are designed and played cannot be taken for granted. Sometimes, there are real-world "games" at play inside and around professional wargames—namely, additional or ulterior motives are at work.

This chapter considers decision-making about cyber wargames and gameplay through the lens of individual cognition and small group psychology. I argue that both warrant serious consideration when creating research wargames, and when playing them. There is an understandable temptation to overemphasize the technical aspects of cyber wargames, given the technical features of cyberspace. But the significance of the cognitive, psychological, and thus human factors involved should not be discounted or ignored by game designers or players, nor by the analysts and educators interested in this tool kit.

To illustrate these social conditions and dynamics, I compare two related but distinct cyber wargames about critical infrastructure: the 2017 *Navy–Private Sector Critical Infrastructure War Game* and the 2019 *Defend Forward: Critical Infrastructure War Game*.[1] On their face, both games had similar research objectives. Both were designed by essentially the same team. And both were played by similar kinds of players. Despite their similarities, however, these two games resulted in very different outcomes. These differences are difficult to explain without considering the cognitive and psychological factors at work for both the design team and the players involved.

By all accounts, the 2017 *Navy–Private Sector Critical Infrastructure War Game* was an analytical success (for further details, see chapter 3). Inspired by recent attacks on critical infrastructure, the research questions were designed to inform future cyber defense policies, and the answers provided by the game were instructive.[2] In contrast, while the 2019 *Defend Forward: Critical Infrastructure War Game* seemed poised for similar success, it fell short from the beginning of the design process.

Rather than focus on the cyber content of these games, I discuss their unspoken social content and context. First, I consider a couple of prominent biases in individual cognition. Second, I provide a similar overview of key concepts in small group dynamics, commenting on the analytical traction that wargaming—cyber or otherwise—can provide. Third, I describe how several of these factors interacted during two cyber wargames about critical infrastructure. Fourth and finally, I argue that explicitly addressing individual cognition and group dynamics can benefit the practice of wargaming during design, analysis, and play.

INDIVIDUAL COGNITION AND GROUP DYNAMICS

Early research on individual decision-making focused on why people so often make mistakes, or at least decisions against their self-interests, when they are faced with uncertainty.[3] The mental shortcuts that we use—for better or worse—to help simplify our thinking and decision-making provided an important answer.[4] This was significant work (some of which was funded by the military). It has proved pivotal not only in psychology but also in almost every field of social science.[5]

Daniel Kahneman and Amos Tversky outlined several of these mental shortcuts, such as representativeness and availability.[6] As the field of research grew, so too did the range of different mental shortcuts that were theorized and studied.[7] Many of the biases, fallacies, and heuristics that people use as shortcuts to simplify thinking and decision-making have been studied by treating the individual person as the relevant unit of analysis. However, there is also considerable research into individual cognition *in the context of* small groups.[8] I briefly consider how a handful of cognitive biases affect and interact with decision-making by the small groups commonly found in wargaming.

Cognitive Biases, Fallacies, and Heuristics

People sometimes see, hear, and think what they want to (or what they previously did), regardless of new and even disconfirming evidence around them. Selectively disregarding evidence that opposes one's belief while holding supportive evidence in higher regard is known as confirmation bias.[9] For example, in one of many experiments on confirmation bias, test subjects were first asked if they supported or opposed capital punishment; then they were presented with two studies about the death penalty—one defending it and one discrediting it—as a deterrent against crime.[10] Most people found the study that supported their prior beliefs to be most convincing. We can imagine analogous outcomes with regard to cybersecurity. For instance, if you already believe that cyberspace has revolutionized warfare, then you may discount or dismiss evidence indicating that it is not important for battlefield outcomes, or vice versa.

Confirmation bias can reinforce the availability heuristic, whereby the ease of recalling prior events can make us think that similar events are likely to occur again.[11] Preparing to "fight the last war" is one common example in wargaming and military planning for real life. Similarly, a recent cyberattack against the electricity grid might lead us to expect that more of the same sorts of attacks are forthcoming. The absence of personal, lived experiences can conversely cause us to discredit other peoples' experiences. A lack of cyberattacks on your electricity grid to date does not necessarily exempt or protect you from such an attack in the future.[12] Availability can drive our expectations and decisions more than statistically significant empirical evidence per se.

Our thinking at any given moment does not occur only in that moment. Nor does it occur in a social vacuum. Granted, not all these quirks in individual cognition are at work all the time, or to the same degree in different people. But when people come together to form groups, their individual quirks can inform group dynamics, as well as be informed by them, constructing their relationships within the group and the behavior of the group as a whole.

Small Group Dynamics

In war, tactical success does not simply add up to operational success or strategic victory. Similarly, individual cognition does not simply aggregate into group decision-making.[13] When people get together, sometimes the whole is greater—or at least different—than the sum of its parts.

Information is of course important for decision-making by individuals and groups alike, even when our perceptions and thinking about that information are biased by confirmation and availability. All else being equal, common sense may suggest that groups can perceive and process more information than an individual and, therefore, that groups will make better-informed decisions. However, strong evidence refutes this notion.[14] For instance, individual members of groups sometimes withhold information. This is known as a "hidden profile."[15] Imagine a small company that is trying to decide whether to transition its information and operational technologies to cloud-based services, but their engineer does not tell the chief executive officer that bandwidth is severely constrained at remote endpoints critical to the business. When group members withhold relevant information, the group cannot make fully informed decisions.[16]

There are many reasons why individuals withhold information.[17] Similarly, there are many other reasons why groups often struggle to make good decisions. At one extreme is groupthink, when groups suffer from excessive cohesion that stifles critical thinking. In this case, members of the group "avoid being too harsh in their judgments of their leaders' or colleagues' ideas," thereby "adopting a soft line of criticism, even in their own thinking."[18] Groupthink can occur when people value the group—and their membership or status in it—above other outcomes. As a consequence, they consciously or unconsciously conform to what seems like the most agreeable stance, nonconforming information notwithstanding.[19] Similarly, people in groups are more likely to discuss commonly shared information over information held only by individual members.[20] The failed Bay of Pigs invasion in 1961 is an often-cited case of how groupthink can have a negative impact on foreign policy.[21] It is easy to imagine roughly analogous cases of groupthink in the history of cyber policy (both in the public and private sectors). Although it is understandable why these dynamics arise, their effects sometimes make groups dumber than necessary.

Just as excess cohesion can be problematic, so too can be fragmentation or insufficient group cohesion. When groups of people fragment or fail to gel, individual factors almost inevitably become more prominent, both as potential causes and effects. Individual motivations can drive as well as mirror or reflect collective motivations inside weak social groups.[22]

For instance, a person's willingness to understand a given problem through discussion and information gathering is called epistemic motivation.[23] While internal drivers can motivate individual group members, epistemic motivation can also be amplified if not generated inside groups (especially those held accountable for how they reach decisions).[24] Other motives can interact or override epistemic motivation. To the extent that individual group members are prosocial (i.e., concerned with the group outcome), the group's epistemic motivation may increase, at least to a point. However, to the extent that individual group members are more proself (i.e., concerned primarily with their own outcomes), then their group's epistemic motivation may decrease.[25]

Not surprisingly, less-cohesive groups are less likely to engage in thorough discussion about problem solving and decision-making.[26] After all, problem solving and decision-making are work and, in groups tasked with collective action, proself members may try to free-ride on everyone else. Free riders seek to benefit from groupwork without contributing themselves.[27] Free riding can spread if people fear "being a sucker" by working when others are not; in the extreme, such groups cannot work together as a team. The line between strong group cohesion that generates robust decision-making and excessive cohesion that leads to groupthink can be thin.

Many scholars have investigated what the strengths and weaknesses of group decision-making imply for national security and foreign policy.[28] National security strategies and foreign policies are typically created and enacted by small groups, which makes it important to understand how they make decisions. Was the National Security Council influenced by groupthink about ransomware after the Colonial Pipeline hack in 2021, for example? What were the members' motives? How was information received, processed, and shared with the president and the national security bureaucracy?

These questions present potential problems for methodological individualism in social scientific research. Said another way, studying individual people may not capture the relational phenomena that emerge when we act and interact as social groups. Fortunately, wargames are appealing because they can provide analytical traction on decision-making by small groups. If designed appropriately, they can likewise provide insight into social relationships, such as the interplay between individual expertise on cybersecurity and the implications for group behavior. As in real life, evidence from cyber wargames suggests that these dynamics can be consequential.

ILLUSTRATIVE EVIDENCE FROM CYBER WARGAMES ABOUT CRITICAL INFRASTRUCTURE

The scholarship on individual cognition and group dynamics in wargames is growing but still incomplete.[29] For example, Stephen Downes-Martin charts the effects of ignoring how players and adjudicators think and feel during wargames (e.g., overconfidence or depression). He also identifies ways to avoid bad outcomes, such as treating adjudicators *as* players.[30] While not couched in terms of wargaming, Rose McDermott examines what emotions imply for policy decisions about cyber warfare, which could be extended to test in or improve cyber wargames.[31]

More work is needed. In the next subsections, I examine some of the key concepts discussed above—specifically, the availability heuristic, groupthink, and motivation. I do so in the context of two critical infrastructure cyber wargames.

Take One: The 2017 Navy–Private Sector Critical Infrastructure War Game

Cyberattacks against critical infrastructure are a hot topic. Critical infrastructure refers to vital systems and assets—physical or virtual—that "would have a debilitating impact" on national security if they were incapacitated or destroyed.[32] Prominent cyberattacks against critical infrastructure include the

2015 BlackEnergy attack on Ukraine's power grid, which was the first time a cyberattack is known to have targeted a power grid.[33] In addition, the Wannacry ransomware attack in May 2017 adversely affected telecommunication in Spain, hospitals across the United Kingdom, and computer systems in more than a hundred other countries.[34] Shortly thereafter, the malware known as NotPetya disrupted critical infrastructure for transportation, as well as financial and public services in Ukraine and around the world.[35]

In 2017, my colleagues and I at the Naval War College were approached by a sponsor who wanted to use wargaming to analyze the implications of these kinds of cyberattacks. The notion of cyber wargaming for critical infrastructure was not new. For example, *GridEx* had been running since 2011, as described in chapter 10. Moreover, in 2006, the Department of Homeland Security sponsored Cyber Storm, "the first Government-led, full-scale, cyber security exercise of its kind," which also simulated cyberattacks against critical infrastructure.[36]

Building on this history, the 2017 *Navy–Private Sector Critical Infrastructure War Game* brought local and federal government officials together with representatives from the private sector. Our analytic objective was to study how individuals and groups in these different positions or roles interacted with each other during significant cyberattacks. Specifically, we wanted to determine the threshold that cyberattacks against private critical infrastructure must reach to be considered a national security concern and, in turn, require government intervention.[37]

To answer this question, we designed a cyber wargame, as summarized in chapter 3.[38] Given that several cyberattacks against critical infrastructure had been in the news at that time, the availability heuristic was probably quite influential in this game. Although availability can reinforce confirmation biases, we observed players at least attempting to explicitly rationalize their responses' cyber actions and effects in the wargame, which may have minimizing overt effects.[39] For example, when the fictional Financial Services Team started to receive incident reports of cyberattacks against its networks, it discussed appropriate solutions as a group, listened to each team member's proposed response, and took a majority vote when different solutions were proposed. The range of different proposals raised seemed to indicate that individual team members were sharing relevant information, thus reducing the risks stemming from potentially hidden profiles.

Debates and majority votes also suggest some resistance to groupthink in this instance. Again, the range of different proposals raised indicates that, while members of a team, these players did not appear obligated to conform to the most agreeable stance. While group cohesion was strong, it was not excessive, and this balance allowed this group to thoughtfully consider each problem they faced—an indicator of high epistemic motivation.

That said, even when epistemic motivation appears high, groupthink is still possible. For example, the Energy Services Team also took internal votes on how to handle the crises it faced in gameplay. However, in this case, a unanimous consensus was more common. While unanimous votes are not proof positive of groupthink, their repetition over time may indicate the emergence of a social norm that quieted or quelled potential—and potentially constructive—dissent on this team.

Individual motivations are complex and difficult if not impossible to observe. However, given that we observed active information sharing and consultation both inside and across the teams in this wargame, the players' epistemic motivation was arguably high. To the extent that most players were also interested in the overall outcome for their group, the same can probably be said about their prosocial motivation. I witnessed similar behavior in the design team and in interactions with the sponsor.[40] Therefore, while this game was not designed to observe small group dynamics, it serves as a useful case study for effective decision-making in future games.

Take Two: The 2019 Critical Infrastructure War Game

Alas, some wargames illustrate how cognitive biases and small group dynamics can have a negative impact on research outcomes and players' experiences. There are times when the motivations of the sponsor, the design team, and the players do not align particularly well. Individual motivations can interact with small group dynamics in unconstructive ways. Even when clashing motivations are the root cause of difficulty, group psychology can further compound the problems.

For instance, when the same sponsor requested another cyber wargame on critical infrastructure in 2019, our design team was initially eager to examine some of the important research questions raised by the previous game. However, our research questions differed somewhat from the sponsor's interests. Consistent with the availability heuristic, we eventually agreed to examine questions about cyber operations that related to recent changes in military strategy, specifically, how this new strategy might affect critical infrastructure.[41] It is possible that our diverging underlying interests led to a less cohesive group, with all that entails for wargame design.

Limited group cohesion was not unique to the design team. Indeed, when it came to gameplay, several of the players' teams also exhibited signs of weak cohesion. For example, some teams randomly assigned roles among their players, whereas other players seemed to take the role they wanted without waiting for approval from their team.[42] I observed that groups with randomly assigned roles seemed to conduct more thorough decision-making, suggesting higher degrees of shared epistemic motivation, compared with groups where the individuals self-selected their roles.

This observation was further supported, with some drama, when one player stormed out of the room in frustration. They never returned to the game. Apparently, their frustration partially stemmed from a perceived lack of respect for their real-life professional status compared with the responsibilities associated with their role in the game—a role they had self-selected.[43] After this player's departure, the team members who remained went through the motions to complete the wargame, but there was little group discussion or coordination.

Again, it is almost impossible to observe a person's true motives. But the observed behavior of the player who stormed out and the subsequent effect on their team suggests that this individual was more concerned with themself than their team. Proself individuals hold noncooperative attitudes. Including a proself individual in an otherwise-cooperative (i.e., prosocial) group can cause other members of the group to also adopt noncooperative attitudes, decreasing their willingness to address shared problems through collaborative discussion (i.e., low epistemic motivation).[44]

Individuals on other teams also commented on noticeably low group engagement. Some indicated that they felt excluded or ignored when trying to share their opinion. Perhaps other members of their teams were only interested in information that supported their prior beliefs, consistent with confirmation biases. It is also possible that other players were most interested in maximizing their personal gain (e.g., professional networking outside the game for relationships and status in real life), rather than engagement through gameplay, cooperation, and information sharing on their team.[45]

As in 2017, the 2019 wargame was not designed to observe these small group dynamics. Therefore, the suggestions given above are simply hypotheses. Nevertheless, this experience still served as a useful window into how wargames can fall short when group dynamics and individual cognition are not considered.

CONCLUSION

Wargames do not occur in a bubble, but they can control for some contextual factors to explore how decisions are made.[46] On one hand, the ability to control information and examine group decision-making is one of the most powerful advantages of wargaming as a research method.[47] On the other hand, the potential importance of these same factors means that ignoring them can frustrate game design, play, and analysis.

By my account, the smooth functioning of the 2017 wargame and the complications with the 2019 wargame illustrate how often-ignored social aspects of wargaming are particularly relevant for cyber wargaming. The good news, for analysts and game designers, is that small group dynamics offer exciting opportunities for future research using this very tool. For example, it is reasonable to hypothesize that robust information sharing within groups can reduce groupthink, confirmation bias, and hidden profiles. Wargame facilitators often implicitly take some of these aspects into account when they lead discussions with players. Following suit with experimental research to test hypotheses like this would strengthen our understanding of group dynamics and decision-making, as well as our ability to build better wargames.

Players can also benefit by taking cognitive biases and group dynamics into account. No one has perfect knowledge. Nor can any group or team guarantee to support each member's mental and emotional state at every point in gameplay. Simulated competition and conflict can be stressful (as in real life), even when that stress is healthy and constructive. As emphasized in the literature on teaching and learning, safely simulating stress for personal growth and professional development is one reason why wargaming works for training and education.[48]

To make the most of the experience, I recommend that players consider their own biases and motivations. Ask yourself, "what would make a successful wargame for me?" Are you trying to learn, for example, or impress your boss? Whatever the case may be, think about how to make the most of gameplay, especially when you must work to constructively reconcile your motives with those of your teammates and other players.

As argued throughout this book, a wargame's objectives are paramount to everything else. Why various explicit and implicit objectives are pursued or ignored—by design teams, sponsors, players, and adjudicators—should not be overlooked. As individuals and groups, our motives and biases can have a profound impact in and through cyberspace, just as they do everywhere else.

NOTES

1. Jacquelyn Schneider, Benjamin Schechter, and Rachael Shaffer, *Navy–Private Sector Critical Infrastructure War Game 2017 Game Report* (Newport: US Naval War College, 2017); Cyber & Innovation Policy Institute (CIPI) Staff, *Defend Forward: Critical Infrastructure War Game 2019 Game Report* (Newport: US Naval War College, 2019).

2. Kim Zetter, "Inside the Cunning, Unprecedented Hack of Ukraine's Power Grid," *Wired*, March 3, 2016, www.wired.com/2016/03/inside-cunning-unprecedented-hack-ukraines-power-grid/; "What Is WannaCry Ransomware?" Kaspersky, no date, https://usa.kaspersky.com/resource-center/threats/ransomware-wannacry; Brian Krebs, "'Petya' Ransomware Outbreak Goes Global," *Krebs on Security*, June 27, 2017, https://krebsonsecurity.com/2017/06/petya-ransomware-outbreak-goes-global/; "What Is Petya and NotPetya Ransomware?" McAfee, no date, www.mcafee.com/enterprise/en-us/security-awareness/ransomware/petya.html.

3. Among others, see Daniel Kahneman and Amos Tversky, "Subjective Probability: A Judgment of Representativeness," *Cognitive Psychology* 3, no. 3 (1972): 430–54; Amos Tversky and Daniel Kahneman, "Availability: A Heuristic for Judging Frequency and Probability," *Cognitive Psychology* 5, no. 2 (1973): 207–32; Amos Tversky and Daniel Kahneman, "Judgment under Uncertainty: Heuristics and Biases," *Science* 185, no. 4157 (1974): 1124–31; and Daniel Kahneman and Amos Tversky, "Prospect Theory: An Analysis of Decision under Risk," *Econometrica* 47, no. 2 (1979): 263–92.

4. Along with Amos Tversky and Daniel Kahneman, another pioneer in this field was Herbert Simon. See Herbert A. Simon, *Administrative Behavior: A Study of Decision-Making Processes in Administrative Organization*, 2nd ed. (New York: Macmillan, 1957).

5. Daniel Kahneman and Amos Tversky, *Intuitive Prediction: Biases and Corrective Procedures* (Eugene, OR: Decision Research, 1977); Gerd Gigerenzer and Wolfgang Gaissmaier, "Heuristic Decision Making," *Annual Review of Psychology* 62, no. 1 (2011): 454.

6. Daniel Kahneman and Amos Tversky, "Subjective Probability: A Judgment of Representativeness," *Cognitive Psychology* 3, no. 3 (1972): 430–54; Tversky and Kahneman, "Availability."

7. The Decision Lab, a consulting firm that uses applied behavioral science to inform its research, has identified almost a hundred "relevant biases" that have an impact on human decision-making. Decision Lab, "Cognitive Biases," no date, https://thedecisionlab.com/biases.

8. Marceline B. R. Kroon, Paul't Hart, and Dik van Kreveld, "Managing Group Decision Making Processes: Individual versus Collective Accountability and Groupthink," *International Journal of Conflict Management* 2, no. 2 (1991): 91–115; Attila Ambrus, Ben Greiner, and Parag A. Pathak, "How Individual Preferences Are Aggregated in Groups: An Experimental Study," *Journal of Public Economics* 129 (September 2015): 1–13; Norbert L. Kerr, Robert J. MacCoun, and Geoffrey P. Kramer, "Bias in Judgment: Comparing Individuals and Groups," *Psychological Review* 103, no. 4 (1996): 687–719; Tatsuya Kameda and James H. Davis, "The Function of the Reference Point in Individual and Group Risk Decision Making," *Organizational Behavior and Human Decision Processes* 46, no. 1 (1990): 55–76.

9. Raymond S. Nickerson, "Confirmation Bias: A Ubiquitous Phenomenon in Many Guises," *Review of General Psychology* 2, no. 2 (1998): 175.

10. Charles G. Lord, Lee Ross, and Mark R. Lepper, "Biased Assimilation and Attitude Polarization: The Effects of Prior Theories on Subsequently Considered Evidence," *Journal of Personality and Social Psychology* 37, no. 11 (1979): 2098–2109.

11. Amos Tversky and Daniel Kahneman, "Availability: A Heuristic for Judging Frequency and Probability," in *Judgment under Uncertainty: Heuristics and Biases*, edited by Daniel Kahneman, Paul Slovic, and Amos Tversky (Cambridge: Cambridge University Press, 1982), 178.

12. Shelley E. Taylor, "The Availability Bias in Social Perception and Interaction," in *Judgment under Uncertainty: Heuristics and Biases*, edited by Daniel Kahneman, Paul Slovic, and Amos Tversky (Cambridge: Cambridge University Press, 1982), 190–200; Philomena Essed, *Understanding Everyday Racism: An Interdisciplinary Theory* (Newbury Park, CA: Sage, 1991), 70; Greta McMorris, "Critical Race Theory, Cognitive Psychology, and the Social Meaning of Race: Why Individualism Will Not Solve Racism," *UMKC Law Review* 67, no. 4 (Summer 1999): 715.

13. Ambrus, Greiner, and Pathak, "How Individual Preferences Are Aggregated."
14. E.g., see Garold Stasscr and William Titus, "Hidden Profiles: A Brief History," *Psychological Inquiry* 14, nos. 3–4 (2003): 304–13; John P. Lightle, John H. Kagel, and Hal R. Arkes, "Information Exchange in Group Decision Making: The Hidden Profile Problem Reconsidered," *Management Science* 55, no. 4 (April 2009): 568–81; Gwen M. Wittenbaum, Andrea B. Hollingshead, and Isabel C. Botero, "From Cooperative to Motivated Information Sharing in Groups: Moving Beyond the Hidden Profiles Paradigm," *Communications Monograph* 71, no. 3 (2004): 286–310; Carsten K. W. De Dreu, Bernard A. Nijstad, and Daan van Knippenberg, "Motivated Information Processing in Group Judgment and Decision Making," *Personality and Social Psychology Review* 12, no. 1 (2008): 22–49; Michael Horowitz et al., "What Makes Foreign Policy Teams Tick: Explaining Variation in Group Performance and Geopolitical Forecasting," *Journal of Politics* 81, no. 4 (2019): 1388–1404.
15. Garold Stasser and William Titus, "Pooling of Unshared Information in Group Decision Making: Biased Information Sampling during Discussion," *Journal of Personality and Social Psychology* 48, no. 6 (1985): 1467–78.
16. Gwen M. Wittenbaum and Garold Stasser, "Management of Information in Small Groups," in *What's Social about Social Cognition? Research on Socially Shared Cognition in Small Groups*, edited by Judith L. Nye and Aaron M. Brower (Thousand Oaks, CA: Sage, 1996), 3–28.
17. For a review, see De Dreu, Nijstad, and van Knippenberg, "Motivated Information Processing."
18. Irving L. Janis, "Groupthink," *Psychology Today* 5, no. 6 (November 1971): 84. Among others, see Irving L. Janis, *Victims of Groupthink: A Psychological Study of Foreign-Policy Decisions and Fiascoes* (Boston: Houghton Mifflin, 1972).
19. Robert B. Zajonc, "The Process of Cognitive Tuning in Communication," *Journal of Abnormal and Social Psychology* 61, no. 2 (1960): 159–67; Sidney Verba, *Small Groups and Political Behavior: A Study of Leadership* (Princeton, NJ: Princeton University Press, 1961); Roby D. Robertson, "Small Group Decision Making: The Uncertain Role of Information in Reducing Uncertainty," *Political Behavior* 2, no. 2 (1980): 163–88.
20. Stasser and Titus, "Pooling of Unshared Information"; De Dreu, Nijstad, and van Knippenberg, "Motivated Information Processing"; Wittenbaum, Hollingshead, and Botero, "From Cooperative to Motivated Information."
21. Janis, "Victims of Groupthink."
22. For reviews, see De Dreu, Nijstad, and van Knippenberg, "Motivated Information Processing"; and Lotte Scholten, Daan van Knippenberg, Bernard A. Nijstad, and Carsten K. W. De Dreu, "Motivated Information Processing and Group Decision-Making: Effects of Process Accountability on Information Processing and Decision Quality," *Journal of Experimental Social Psychology* 43, no. 4 (2007): 539–52.
23. De Dreu, Nijstad, and van Knippenberg, "Motivated Information Processing," 233.
24. Karen Siegel-Jacobs and J. Frank Yates, "Effects of Procedural and Outcome Accountability on Judgment Quality," *Organizational Behavior and Human Decision Process* 65, no. 1 (1996): 1–17; Scholten et al., "Motivated Information Processing," 541.
25. De Dreu, Nijstad, and van Knippenberg, "Motivated Information Processing," 38.
26. Charles R. Evans and Kenneth L. Dion, "Group Cohesion and Performance: A Meta-Analysis," *Small Group Research* 22, no. 2 (May 1991): 175–86.

27. Robert Albanese, and David D. van Fleet, "Rational Behavior in Groups: The Free-Riding Tendency," *Academy of Management Review* 10, no. 2 (1985): 244–55, https://doi.org/10.2307/257966.

28. E.g., see Janice Gross Stein, "Building Politics into Psychology: The Misperception of Threat," *Political Psychology* 9, no. 2 (1988): 245–71; Alexander L. George, "The Two Cultures of Academia and Policy-Making: Bridging the Gap," in "Political Psychology and the Work of Alexander L. George," special issue, *Political Psychology* 15, no. 1 (March 1994): 143–72; Jack S. Levy, "Prospect Theory, Rational Choice, and International Relations," *International Studies Quarterly* 41, no. 1 (1997): 87–112; McDermott, *Risk-Taking*; Alex Mintz, "Behavioral IR as a Subfield of International Relations," *International Studies Review* 9, no. 1 (2009): 157–72; Mark Schafer and Scott Crichlow, *Groupthink versus High-Quality Decision Making in International Relations* (New York: Columbia University Press, 2010); Robert Jervis, *Perception and Misperception in International Relations* (Princeton, NJ: Princeton University Press, 2017); Janice Gross Stein, "The Micro-Foundations of International Relations Theory: Psychology and Behavioral Economics," *International Organization* 71, no. 1 (2017): S249–S263, https://doi.org/10.1017/S0020818316000436; and Horowitz et al., "What Makes Foreign Policy Teams Tick."

29. Stephen Downes-Martin, "Adjudication: The *Diabolus in Machina* of War Gaming," *Naval War College Review* 66, no. 3 (2013): 67–80, 165; Shay Hershkovitz, "Wargame Business: Wargames in Military and Corporate Settings," *Naval War College Review* 7, no. 2 (2019): 67–82; Walter W. Kulzy, "(Design) Thinking Through Strategic-Level Wargames for Innovative Solutions," *Military Operations Research Society* 52, no. 2 (2019): 56–62.

30. Downes-Martin, "Adjudication," 165.

31. Rose McDermott, "Some Emotional Considerations in Cyber Conflict," *Journal of Cyber Policy* 4, no. 3 (2019): 309–25; Benjamin Dean and Rose McDermott, "A Research Agenda to Improve Decision Making in Cyber Security Policy," *Penn State Journal of Law & International Affairs* 5, no. 1 (2017): 30–71.

32. Critical Infrastructure Act, 42 U. Code § 5195c, 2001.

33. Zetter, "Inside the Cunning." It is believed that another cyberattack was responsible for the 2016 power outages in Ukraine. See Pavel Polityuk, Oleg Vukmanovic, and Stephen Jewkes, "Ukraine's Power Outage Was a Cyber Attack: Ukrenergo," Reuters, January 18, 2017, www.reuters.com/article/us-ukraine-cyber-attack-energy/ukraines-power-outage-was-a-cyber-attack-ukrenergo-idUSKBN1521BA.

34. "What Is WannaCry Ransomware?"

35. Krebs, "'Petya' Ransomware"; "What Is Petya and NotPetya Ransomware?"

36. US Department of Homeland Security, National Cyber Security Division, *Cyber Storm Exercise Report* (Washington, DC: US Department of Homeland Security, 2006).

37. Schneider, Schechter, and Shaffer, *Navy–Private Sector*, 7.

38. For more information about the mechanics of the game itself, see the game report: Jacquelyn Schneider, Benjamin Schechter, and Rachael Shaffer, *Navy–Private Sector Critical Infrastructure War Game 2017 Game Report* (Newport: US Naval War College, 2017).

39. Deanna Kuhn, "Thinking as Argument," *Harvard Educational Review* 62, no. 2 (1992): 155.

40. Schneider, Schechter, Shaffer, *Navy–Private Sector*, 21.

41. For more information about the mechanics of the game itself, see the game report: Cyber & Innovation Policy Institute (CIPI) Staff, *Defend Forward: Critical Infrastructure War Game 2019 Game Report* (Newport: US Naval War College, 2019), 2.

42. This was not a variable we were testing for at the time of the wargame, so it is recalled to the best of my memory. Similiary, why certain groups elected for difference selection processes was based on individual personalities.

43. We asked players to not only assign the roles among their team members but to also assign responsibilities. CIPI Staff, *Defend Forward*, 4.

44. Harold H. Kelley and Anthony J. Stahelski, "Social Interaction Basis of Cooperators' and Competitors' Beliefs about Others," *Journal of Personality and Social Psychology* 16 (1970): 66–91; Paul A. M. van Lange, "Confidence in Expectations: A Test of the Triangle Hypothesis," *European Journal of Personality* 6, no. 5 (1992): 371–79; Wolfgang Steinel and Carsten K. W. De Dreu, "Social Motives and Strategic Misrepresentation in Social Decision Making," *Journal of Personality and Social Psychology* 86, no. 3 (2004): 419–34; Femke Ten Velden, Bianca Beersma, and Carsten K. W. De Dreu, "Majority and Minority Influence in Group Negotiation: The Moderating Effects of Social Motivation and Decision Rules," *Journal of Applied Psychology* 92, no. 1 (2007): 259–68.

45. De Dreu, Nijstad, and van Knippenberg, "Motivated Information Processing," 42; Kelley and Stahelski, "Social Interaction Basis"; Dale T. Miller and John G. Holmes, "The Role of Situational Restrictiveness on Self-Fulfilling Prophecies: A Theoretical and Empirical Extension of Kelley and Stahelski's Triangle Hypothesis," *Journal of Personality and Social Psychology* 31, no. 4 (1975): 661–73.

46. Shawn Burns, *War Gamers' Handbook: A Guide for Professional War Games* (Newport: US Naval War College, 2013), 42.

47. Benjamin Schechter, Jacquelyn Schneider, and Rachael Shaffer, "Wargaming as a Methodology: The International Crisis Wargame and Experimental Wargaming," *Simulation & Gaming* 52, no. 4 (August 2021): 513–26.

48. David T. Long, and Christopher M. Mulch, *Interactive Wargaming Cyberwar: 2025* (Monterey, CA: Naval Postgraduate School, 2017).

PART II

EDUCATIONAL GAMES

8

PLAYFUL LEARNING ABOUT CYBERSECURITY

ANDREAS HAGGMAN

Despite shared nomenclature, wargames and play are not easy bedfellows.[1] Professional wargames typically seek legitimacy by appearing to be "serious," which is usually assumed to be the antithesis of play. This dichotomy is misguided, however. In many cases, playfulness can help make wargaming more rather than less effective.

This chapter reconciles wargaming with play and outlines the impact of playfulness on one educational wargame about cybersecurity. Compared with more passive methods for teaching and learning, wargaming is often touted as an engaging activity for effective education.[2] The engaging aspects of this game-based learning—similar to but somewhat distinct from the related concept of gamification—stem in part from the active participation and interaction of players that serve "to foster cognitive engagement of the learner with the learning mechanic."[3] I argue that engagement is also generated through the sense of fun that comes with playful behavior.

Playfulness is not a foreign concept to cybersecurity. For example, old-school hackers like Robert T. Morris and Kevin Mitnick claim that their exploits were driven by curiosity rather than more nefarious motives.[4] While this justification for illegal activity is self-serving, some hackers may well be most interested in exploring networked computer systems and chasing the thrill of evading detection, rather than breaking or stealing things per se.[5] The practice of hiding and hunting for "easter eggs" in video games and other software can also reward curious users in the spirit of exploration.[6] In a sense, this chapter connects some of the playfulness already present in cybersecurity to the seemingly more serious practice of cyber wargaming.

The chapter has three main sections. First, I interrogate the literature on wargaming, the study of play, and pedagogy to build theoretical support for reconciling serious games and playful learning. Second, I outline the design of an original cybersecurity strategy game, along with how I used this game for academic research on education. Third and finally, I describe my findings: most notably novel insight into the role of humor, which is almost entirely absent from the extant literature on wargaming.

THE IMPORTANCE OF BEING PLAYFUL

The fraught relationship between wargaming and playfulness is symbolized in part by a subtle schism between the terminology preferred by different wargaming communities. The separated term "war gaming" is preferred by some professionals, whereas the closed-up term "wargaming" is more popular among hobbyists. Peter Perla suggests that the physical separation of these two words resonates with professionals who seek to distance themselves from the jovial "gaming" aspects of wargaming.[7] Be that as it may, there are historical, ludological (ludology being the study of play), and pedagogical reasons why such distancing ought to be avoided.

Historical Significance

Despite the obvious seriousness of war as a literal life and death matter, combat has long been linked with playfulness as well. Roger Caillois, a notable scholar of ludology, characterizes play as something that is voluntary (i.e., not obligatory). These characteristics render play an "attractive and joyous quality as a diversion."[8] Ironically, perhaps, war is sometimes rendered in similar ways.

For example, the ancient Greeks seemed to consider war as something of a sport, replete with glory and rivalry.[9] In medieval Europe, warfare was closely entwined with ideas of chivalry and thus games for social status. Battle skills associated with honor and courage were showcased in tourneys, holding martial ritual hand in hand with entertainment.[10] Such sporting events suggest that, in the past, war and play were not far apart in people's minds. As Thomas B. Allen suggests, "the idea that war itself could be waged like a game . . . is well chronicled through the age of chivalry."[11] Indeed, similar sentiments persist into modern times, when we consider the language of sports. We speak of teams "going to battle," players "in the trenches," and a proficient leader of a soccer team might even be referred to as a "midfield general."[12]

Today, wars maintain their serious causes, conduct, and deadly consequences. But they also retain oddly if not disturbingly cathartic elements for individuals and social groups, beyond mere analogies or metaphors about war and sport. As the philosopher John Gray argues, wars have "become another entertainment."[13] Indeed, James der Derian's extension of the classic military-industrial complex into "the military-industrial-media-entertainment network" illustrates how thoroughly war pervades our consumption of modern entertainment.[14]

The relationship between war and entertainment will likely resonate with anyone who followed the television coverage of the 1990–91 Gulf War. The extensive and visually elaborate reporting of that conflict led to it being called the "first Nintendo war."[15] More recently, reporting has described the Russian invasion of Ukraine in 2022—accurately or not—as "the first social media war."[16] People in Ukraine have used social media platforms like TikTok to garner international support by streaming images of war live; the Ukrainian government has likewise launched the chatbot-using messaging service Telegram (eVorog), where citizens can report Russian troop movements.[17] These platforms, originally intended for communication and entertainment, have been co-opted to bring the visceral spectacle of war to screens all over the world (often in deeply personal ways). Ignoring the myriad links between war and entertainment would deny us important opportunities to understand the nature and character of war, which ultimately is one aim of wargaming.

Note that entertainment and play are not the same. But they are related. On one hand, Peter G. Stromberg, suggests that entertainment "is always some sort of play," but "not all play is entertainment."[18] On the other hand, it can be posited that play is a sufficient, though not necessary, condition of entertainment. In other words, playfulness can be entertaining (both for players and observers), but entertainment can also come from other sources. In this view, the key difference is that play is active while entertainment can be passive. As an illustration, consider playing a sport versus idly watching a sports event on television.

For the purpose of this chapter, the definition of play provided by Katie Salen and Eric Zimmerman is useful: "Play is free movement within a more rigid structure."[19] Here play is an emergent property, occurring both because of and despite the constraints of the system. It is also useful to adopt Salen and Zimmerman's definition of being playful, which "refers not only to typical play activities but also to the idea of being in a playful state of mind, where the spirit of play is injected in some other action."[20]

Ludology

Ludology—the study of games—can trace its foundations as a scholarly field to the middle of the twentieth century, particularly in social and cultural studies. The Dutch historian Johan Huizinga's 1938 work *Homo Ludens* is seminal in the field, establishing play and games as key components for building culture: "A given magnitude existing before culture itself existed."[21] The French sociologist Roger Caillois's *Man, Play, and Games* of 1958 extended, but also criticized, Huizinga, arguing that the latter's definition of play was "at the same time too broad and too narrow."[22] Caillois preferred to think of play as consisting of six elements: free, separate, uncertain, unproductive, governed by rules, and make-believe.[23] With the rise of computer games, narratology—the study of games through literary theory—focused more on games as texts (i.e., stories with characters, endings, and settings). Narratology gained prominence over ludology, at least until Gonzalo Frasca sought to reconcile the two approaches in 1999. Frasca coined the term "ludology," borrowing from Caillois's delineation between play ("paidea") and game ("ludus").[24] Frasca also placed renewed emphasis on the role of the player: the human playing the game.

Ludology scholarship continues to grow and develop today, partly driven by new gaming technologies that determine how games are played and experienced. Wargaming, for example, is at heart an exercise in human decision-making during competition and conflict. Human decisions under these conditions may in turn say something about the content of our hearts and minds. With its roots in social and cultural studies, as well as Frasca's emphasis on studying players, ludology offers important opportunities to further understand the human elements of war.

Huizinga drew explicit links between war, play, and culture, suggesting that "the innate desire to be first" drives people to conflict.[25] Political science has since provided alternative and more robust explanations for the causes of war and international conflict.[26] But desire, motivation, or willpower is still recognized—along with factors such as capability and skill—as critical to combat. As Ben Connable and colleagues put it: "[The] will to fight has endured as a central theme or at least a strong undercurrent in military theory for centuries."[27] And similar to how the necessities of war are sometimes said to be the proverbial mother of invention, Alfred H. Hausrath argues that gaming "challenges the competitive spirit" and thereby "encourages innovation."[28]

Competition is critical in cyberspace as well. Many aspects of cybersecurity are adversarial, with attackers and defenders jostling for position and control of networked information technologies. Arguably, these contests fuel what may prove to be a perpetual arms race for ever-more-potent offensive and defensive capabilities in this domain.[29]

Wargaming is said to be a safe environment in which to practice inherently dangerous activities. Or, in Peter Perla's words, "glory without gore and defeat without destruction."[30] Similarly, from the ludological perspective in *The Well-Played Game*, Bernie de Koven outlines how players' participation in games is predicated on an atmosphere where players feel comfortable and can take part on their own terms.[31] De Koven's argument resonates with Perla's idea that games provide safe environments, and in these environments players can innovate and compete, as per Hausrath's assertion. It also links to Caillois's conception of games as being both free (players participate if they want) and governed by rules (constraining the game space), which in turn helps us recall Salen and Zimmerman's definition of play as free movement within a more rigid structure.

Pedagogy

In educational settings, the more engaged participants are in a learning activity, the greater the chance that lessons will be received and retained. Engagement can be generated by activities that require active participation, rather than passive absorption, and contain a sense of fun.[32] Indeed, according to Dan Rea and colleagues, fun activities are not only an instrument of learning success but also "learning satisfaction."[33] Reinforcing this point, Bob Stremba and Christian Bisson note that "the idea of using play as a valid form of pedagogy has been substantiated by many scholars."[34] In studies of serious games, students having fun has been linked to "higher interest in the subject matter."[35] Fun also offers opportunities to overcome reluctance to participate in "long, complex and difficult" learning processes, several of which are described in this book.[36]

One complement to playfulness in learning is humor. Humor is perhaps most simply understood as that which "is actually or potentially funny, and the process by which this 'funniness' occurs."[37] According to Jerry Palmer, it is also dependent on a "playful state of mind" shared between humor maker and audience, which means that humor shares a core characteristic of playfulness as defined by Salen and Zimmerman.[38] Humor is a fundamentally human trait; Simon Critchley goes as far as to suggest that Huizinga's homo ludens argument was erroneous, and that instead "we are homo ridens, laughing beings."[39] Academic debate notwithstanding, playfulness and humor appear inextricably linked to the human condition.

Humor has been recognized as an important element in several pedagogical fields.[40] The same is true for ludology.[41] Humor even appears in security studies; as Carol Cohn finds, it can be a prominent tool—be it wry or perverse—for managing if not dulling the gruesome parts of imagining war.[42] However, humor is almost entirely absent from the literature on wargaming. Many foundational books in the field rarely if ever refer to the subject.[43] Perhaps this omission is further evidence of the uneasy relationship between playfulness and professional wargaming. But it is still strange. Studies in adjacent fields such as pedagogy and technology, cited above, show that humor can break down barriers between players and games, and between students and teachers, while increasing participants' engagement. There is good reason to suspect similar benefits for educational wargaming, even if this possibility has been understudied to date.

I have seen first-hand that players' engagement, emotion, entertainment, and humor can serve as vehicles for student learning through cyber wargames. I have also grappled with potential trade-offs, as described in chapter 2, between players' engagement and other values, such as contextual realism. Despite these dilemmas of design, I argue that the educational benefits of engaging play are well worth the effort.

CASE IN POINT: A UK CYBERSECURITY WARGAME

While several chapters in this volume focus on cyber wargames designed and played in the United States, cyber wargaming is not—of course—a uniquely American enterprise. Working "across the pond," I designed an educational wargame about cyber strategy in a European context. The game was created as part of an academic research project to investigate how wargaming might be used for cybersecurity education.[44] The next subsections first outline how I designed the game, and then how I deployed it with various audiences.

The Great [Cyber] Game

The intended purpose of the *Great [Cyber] Game* is to serve as a primer for cybersecurity. In other words, the game is designed as an entry point from which student players can go on to build further knowledge. The target audience is people who understand that cybersecurity is important but have not had a chance to meaningfully engage with its core concepts and terminology. It was originally envisaged that this game would be for senior policymakers. The central game mechanics therefore focus on resource management, which is a playful way to represent the real-life tensions and trade-offs faced by people in such roles.

This wargame was inspired by the UK National Cyber Security Strategy. The United Kingdom published its first strategy for cybersecurity in 2011. That document referenced three main actors: government, businesses, and individuals. While updated versions of the strategy were published in 2016 and 2021, much of the original document remains valid.[45] Starting with these three actors, I disaggregated "government" into a civilian government and a military/intelligence actor; I likewise split "business" into a general industry and a critical infrastructure actor. As illustrated in table 8.1, these five roles constitute the playable actors in the game.

The game has two national teams: the United Kingdom and Russia. Admittedly, Russia is explicitly identified as the antagonist. For example, the role of Russian individuals is caricatured as "online trolls." This characterization was legitimate—even before Russia's full-scale invasion of Ukraine in 2022—after the UK government officially attributed several cyberattacks to Russia.[46] Built-in antagonism also sets gameplay up for competition and conflict, which is part of the point.

The game can be played with two to ten players, with each player controlling one or more actors. In practice, the game works best with six to eight players (i.e., three to four players per national team). Those numbers strike a good balance between creating intriguing team dynamics, giving all players enough tasks to stay engaged, and generating educational conversation. Each actor is

TABLE 8.1. Actors Played in the *Great [Cyber] Game*

United Kingdom	Russia
Civilian government	Civilian government
UK Plc (business and industry)	Energetic Bear
Electorate (individuals)	Online trolls
Government Communications Headquarters (GCHQ)	Special Communication Service (SCS)
UK Energy (critical infrastructure)	Rosenergoatom (Russian nuclear power operator)

assigned unique objectives. These objectives are a mix of defense and offense over the short and long terms. Some objectives are purposefully contradictory, so it is impossible for national teams to achieve them all. The intent is to force players to make decisions about which objectives to pursue, and to reassess these decisions as the game progresses, thereby teaching them about hard choices involved with cyber strategy in real life.

The game is played on a bespoke physical gameboard. All ten actors and key connections between them are represented on this board, as seen in figure 8.1. In addition to the main game board, there are two further components. The first one is an illicit market, represented on an adjoining game board and with a selection of cards. Here teams can buy offensive and defensive cyber capabilities (some with a timed duration and some single-use). The illicit market works through an auction, which allows teams to outbid each other and enter arms races. The second component is a deck of event cards, which represent geopolitical developments that are difficult to anticipate but nonetheless have an impact on the game. Event cards are drawn randomly at the start of each turn. As in real life, it is difficult to predict the future and, while not perfectly realistic, this game element adds chance to the political dynamics at play.

Adding another element of chance, the game also uses dice, which are rolled to determine the success or failure of cyberattacks. Informed by feedback from playtesting, the game uses a regular six-sided die, with high dice rolls indicating a successful cyber action. (While not my original design, these changes align with popular board game mechanics, which are more accessible for players not accustomed to the irregular dice sometimes used in specialized wargames.)

Gameplay revolves around resource management, with players spending limited resources on actions to achieve their objectives. Players' actions include transferring resources between different actors, boosting the vitality (hit points) of an actor to increase their survivability and resilience, bidding on the illicit market, and launching cyberattacks along attack vectors. Teams never have

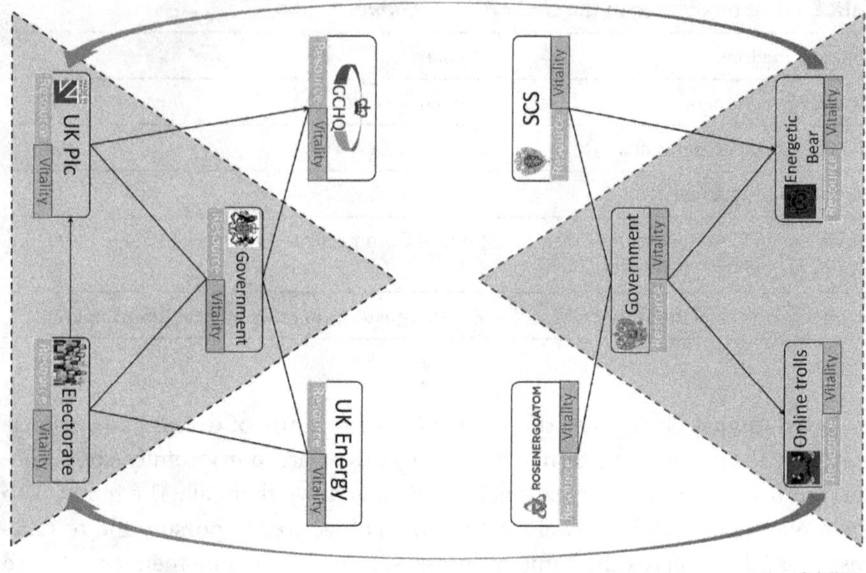

FIGURE 8.1. The *Great [Cyber] Game*—Main Gameboard

enough resources to achieve all objectives. Here again, this is realistic and it also forces players to make trade-offs between their priorities. The game is played for twelve turns that represent the months of a year. After a year of gametime, each national team's victory points—gained for completing offensive and defensive objectives—are tallied up.

Data Collection

During my research, this particular game was played by 259 participants across 33 game sessions. Participants were drawn from a variety of audiences that included the military, government, industry, and academia—often with a mix of different players in any given game session. I ran the game in several different countries (specifically, the United Kingdom, Germany, Sweden, Turkey, Estonia, and Australia), and in several different locations (ranging from university classrooms and company offices to government conference rooms and, on two occasions, local bars). This range of participants and varied settings provided rich qualitative insights into this game's capacity to help players to learn about cybersecurity.

Game sessions were normally conducted in three phases, spanning about 2 hours in total. First, a 10-minute briefing explained the purpose, setup, and rules of the game. Next, the game was played from start to finish, which usually lasted about 90 minutes. Third, the game would close with 20 minutes of free-flowing discussion among the players about their experiences and lessons learned. At any time, participants were welcome to safely inject their own commentary on the game's thematic content, thereby enabling learning and teachable moments for the whole group. Research observations were captured in notes and later analyzed. Several key observations from my data are as follows.

Engagement

In most game sessions, the atmosphere among participants was one of high engagement. In this sense, engagement refers to keeping players interested throughout the game (i.e., limiting players from seeking nongame distractions), and, relatedly, to immersing players in the game experience. The theoretical foundations outlined previously indicate that high engagement, predicated on play and fun, leads to successful pedagogical outcomes. As illustrated in the empirical examples below, players' behaviors can be said to exhibit the "spirit of play" referred to in Salen and Zimmerman's definition of playfulness. Players interacted with the game in playful ways by having fun, reacting to dice rolls, engaging in roleplay, and sharing humor.

FUN

Building on the premise of Rea and colleagues regarding the importance of fun in learning, the reactions of many players suggest that this game was in fact a fun activity. One player, who said they "really, really" (speaker's emphasis) enjoyed the game, exemplified the sentiment of many participants.

Another player offered more critical feedback, arguing that the game was fun for general education but imparted limited new knowledge. This player came from a military and cybersecurity background. Given their subject matter expertise, they may not have been as receptive to the educational content in the game, but they nonetheless appreciated the value of fun in learning. Alternatively, another expert reported that their impression as, "at first, I was like 'what the hell is this?' But after three turns I was like 'hell yeah, let's do this!'" As noted in chapter 1, the practice of cyber wargaming is sensitive to players' background knowledge and experience. How this background plays out can vary, even among experts in the field, and even experts were at least receptive to the idea of having fun and enjoying learning through cyber wargaming.

DICE

The simple game device of using dice contributed to player engagement. Observations of players' reactions to dice rolls indicated a significant degree of emotional investment, with jubilant celebrations and despondent groans meeting each roll. These reactions were not limited to the roller; they included all players around the table. These reactions were especially vociferous over high-stakes rolls, when the dice could potentially determine the outcome of the game. No doubt you are familiar with similar reactions to dice in other settings—a promotional video for this game that captured these dramatics can also be viewed online.[47] Such emotional responses speak to what Caillois calls the "joyous quality" that characterizes gameplay.

It is worth noting that physical dice may be particularly effective in provoking these emotional reactions, the virtual aspects of cyberspace notwithstanding. Marcus Carter and colleagues argue that the feel of dice and the sounds they make give valuable sensory feedback to players.[48] Even digital versions of board games incorporate dice-rolling sounds, emulating some of the sensations and resultant emotions that players attach to rolling dice. Indeed, online gambling sites seek to recreate such sensory stimulation to engage players who are accustomed to the sights and sounds of physical casinos.[49] The very materiality of dice is closely linked to the hands-on rolling action, which may be a factor in the success of this cybersecurity game in creating a high-engagement atmosphere.

ROLEPLAY

Rather than generic red and blue teams or fictional countries, this game was designed around real nations, namely, the United Kingdom and Russia. This approach has contributed to high levels of player engagement, as evidenced by moments of roleplay observed across many game sessions.

Roleplay was prevalent among players on the Russian team. For better or worse, because participants were primarily Western audiences, this roleplay often manifested in terms of national stereotypes. For example, in several game sessions, players on Russian teams joked about sending their less successful teammates to "gulags." One team encouraged its own offensive approach with the phrase "go on, be Putin." The same team also decided against a more defense focused strategy, arguing that it would be uncharacteristic of Russia's approach to cyber operations. Another team joked that, to complete the image, they "need some fur hats." In yet another game (with a completely different audience), one participant turned up in a full Russian military uniform costume, and other participants suggested that a bottle of vodka would help complete the picture.

The UK teams took a somewhat similar and playfully self-critical approach to roleplay, which fortunately served as an implicit check against chauvinism or jingoism. For example, in one session, a player in the role of UK Energy stepped away from the game to address issues in real life. By the time they returned, their team had lost and UK Energy was shattered. They reacted to this by exclaiming, "Oh I don't care, I just want to know if I still have my pension!" Arguably, this response was consistent with roleplay reflecting a—realistically—lackadaisical approach to corporate cybersecurity. Examples such as these illustrate the importance of playfulness in generating a high level of engagement, which can in turn lay the foundation that supports desired learning outcomes.

HUMOR

Like roleplay, it is also worth analyzing the role of humor in gameplay. My game design incorporated humorous elements through selected event cards. A card titled "Clumsy Civil Servant," for example, used an image from the British comedy *Yes, Minister* as a humorous reference. This card referred to misplaced hardware (an all-too-common occurrence in real life). But several players recognized the *Yes, Minister* reference and said as much, which in turn generated further discussion during the game. This humorous element therefore contributed to a higher level of engagement, albeit almost exclusively for British audiences who were familiar with this particular sitcom. More often than not, humor is both context- and culturally dependent; jokes do not always land when translated for different settings and audiences. The quintessential Britishness of *Yes, Minister* exemplifies some of the difficulties of translating humor.[50]

The players themselves were another source of humor, as suggested by the jokes about roleplay discussed above. Its educational utility runs deeper, however. According to Klaus Dodds and Philip Kirby, humor "makes the familiar seem fantastical."[51] By making the familiar fantastical, humor can take important but sometimes mundane realities and elevate them to be more entertaining and engaging.

Serious messages—including lessons about cybersecurity—are often boring. Similar challenges of teaching and learning military doctrine are described in chapter 13. How many times have you heard that you should use strong passwords and not click on suspicious links? These reasonable recommendations often come across as dull and tiresome. Humor, when carefully employed, can be an important tool to introduce serious topics by less serious means. Whereas Cohn describes the gallows humor that defense intellectuals use to cope with the horrors of nuclear war, I find that it may likewise help students cope with the social and technical complexities of cyberspace.[52]

For example, one player drew on their technical knowledge to complain about the opposing team taking too long in their game turn, joking that "they've got some radio frequency manipulator targeting your watch." On a different theme, in reference to fake news influencing the UK electorate, one player quipped, "this is post-Brexit, they have already been influenced." Another player on a Russian team likewise joked that the best way to target the British electorate was to "get at them through the *Sun*," a popular tabloid. These wry comments addressed the very serious role of disinformation, news, and social media in democratic processes linked to cybersecurity.

Other jokes lambasted the budget for GCHQ. Details about this agency's budget are secret, and yet these outbursts were one way for players to express their views without discussing classified information. Here, humor seemed to provide a veil, as well as some degree of catharsis. In a separate discussion about the legality of offensive cyber operations, players acknowledged that the laws of war prohibit targeting civilian infrastructure and yet one participant said, "it's a different question if we're just talking about the lights going off in Moscow." Possibly mundane points were addressed with sharp wit without undermining the playful and fun elements of the game. In this sense, the game encouraged players to make humorous contributions that also served as a learning opportunity.

That said, one potential pitfall of humor is when it crosses the threshold into silliness. Silly behavior and remarks can undermine a wargame, not to mention risk insensitivity and insults that could cause injury as well as disregard for experiential learning. Both humor and silliness depend on context, which can make it difficult to know where the line between them lies. Useful humor requires an ability to "read the room."[53] Because it is not always possible to know the audience ahead of time, or the group dynamics therein (see chapter 7), a prudent approach for designers, teachers, and players is to err on the side of caution. Yes, build humor into the game lightly and allow other participants space to inject their own. But never sacrifice the educational objectives—let alone other players' dignity or fun—for a cheap shot or low blow. Educational wargaming should be an opportunity for all players to learn and grow, not to harden erroneous or simplistic preconceptions.

CONCLUSION

Despite dealing with serious subjects, the playful aspects of wargaming merit consideration. Indeed, for educational wargaming, the dynamics of play and playful behavior by participants can be a core ingredient for success (in partial contrast to analytical wargaming). As argued in this chapter, playfulness generates engagement, which lays the groundwork for participants to make meaningful contributions through gameplay that foster active learning.

In my experience, playfulness is an important factor for achieving learning outcomes. Knowledge about cybersecurity can vary significantly across different audiences. Playful behavior, ranging from emotive reactions and roleplay to humor, can help diverse audiences participate in wargames in ways that build and share knowledge on somewhat common ground.

Playfulness may not be an appropriate vehicle to achieve the desired outcome in all wargames. But all wargames have players. Therefore, understanding how playful behavior works is important for the effective use of this educational and analytical tool. Designers and educators who use these games can harness play as a potent force. Student players can do the same, both for their own benefit and to facilitate teamwork. It is not wrong to say "please have fun."

NOTES

1. This work represents the opinion of the author and in no way represents the opinions or policies of the government of the United Kingdom. Parts of this work were conducted under EPSRC grant EP/K035584/1.
2. Philip Sabin, "Wargaming in Higher Education: Contributions and Challenges," *Arts and Humanities in Higher Education* 14, 4 (2015): 392–48; Johan Elg, "Wargaming in Military Education for Army Officers and Officer Cadets" (doctoral thesis, King's College London, 2017), 182; Graham Longley-Brown, "High-Engagement Wargames," presentation given at Connections UK Conference, September 5, 2015, www.professionalwargaming.co.uk/2016Engagement.pdf.
3. Jan L. Plass, Bruce D. Homer, and Charles K. Kinzer, "Foundations of Game-Based Learning," *Educational Psychologist* 50 (2015): 260; Sebastian Deterding, Dan Dixon, Rilla Khaled, and Lennart Nacke, "From Game Design Elements to Gamefulness: Defining 'Gamification,'" *MindTrek'11*, 2011; Michael Sailer and Lisa Homner, "The Gamification of Learning: A Meta-analysis," *Educational Psychology Review* 32 (2020): 77–112; Steffi Domagk, Ruth N. Schwartz, and Jan L. Plass, "Interactivity in Multimedia Learning: An Integrated Model," *Computers in Human Behavior* 26 (2010): 1024–33.
4. Hilarie Orman, "The Morris Worm: A Fifteen-Year Perspective," *IEEE Security & Privacy* 1 (2003): 35–43; Kevin Mitnick, *Ghost in the Wires: My Adventures as the World's Most Wanted Hacker* (New York: Little, Brown, 2012).
5. Mitnick, *Ghost*.
6. Anastasiia Chuvaieva, "How to Make a Video Game Easter Egg: Legal Tips and Tricks," *Journal of Intellectual Property Law & Practice* 14 (2019): 864–75; Matthew Lakier and Daniel Vogel, "More Than Just Software Surprises: Purposes, Processes and Directions for Software Application Easter Eggs," *Proceedings of the ACM on Human-Computer Interaction* 6 (2022): 1–26.
7. Peter Perla, *The Art of Wargaming: A Guide for Professionals and Hobbyists* (Annapolis, MD: Naval Institute Press, 1990), 2.
8. Roger Caillois, *Man, Play and Games,* translated by Meyer Barash (Urbana: University of Illinois Press, 2001), 9–10.
9. John Gray, *Straw Dogs: Thoughts on Humans and Other Animals* (London: Granta, 2003), 182.
10. John A. Lynn, *Battle: A History of Combat and Culture* (Oxford: Westview Press, 2003), 93.
11. Thomas B. Allen, "The Evolution of Wargaming: From Chessboard to Marine Doom," in *War and Games*, edited by T. J. Cornell and T. B. Allen (Woodbridge, UK: Boydell Press, 2002), 232.
12. Premier Skills English, "Midfielders," https://premierskillsenglish.britishcouncil.org/course-stages/midfielders.
13. Gray, *Straw Dogs*, 183.

14. James der Derian, *Virtuous War: Mapping the Military-Industrial-Media-Entertainment Network*, 2nd ed. (New York: Routledge, 2009).

15. Nina B. Huntemann and Matthew Thomas Payne, "Introduction," in *Joystick Soldiers: The Politics of Play in Military Video Games*, edited by Nina B. Huntemann and Matthew Thomas Payne (Abingdon, UK: Routledge, 2010), 6.

16. Peter Suciu, "Is Russia's Invasion of Ukraine the First Social Media War?" *Forbes*, March 1, 2022, www.forbes.com/sites/petersuciu/2022/03/01/is-russias-invasion-of-ukraine-the-first-social-media-war/?sh=197856eb1c5c.

17. Drew Harwell, "Instead of Consumer Software, Ukraine's Tech Workers Build Apps of War," *Washington Post*, March 24, 2022, www.washingtonpost.com/technology/2022/03/24/ukraine-war-apps-russian-invasion/.

18. Peter G. Stromberg, *Caught in Play: How Entertainment Works on You* (Stanford, CA: Stanford University Press, 2009), 9.

19. Katie Salen and Eric Zimmerman, *Rules of Play: Game Design Fundamentals* (Cambridge, MA: MIT Press, 2004), 304.

20. Salen and Zimmerman, 303.

21. Johan Huizinga, *Homo Ludens: A Study of the Play-Element in Culture* (London: Routledge, 1949), 4.

22. Caillois, *Man, Play, and Games*, 4.

23. Caillois, 9–10.

24. Gonzalo Frasca, "Ludology Meets Narratology. Similitude and Differences between (Video)games and Narrative," *Parnasso* 3 (1999), 365–71.

25. Huizinga, *Homo*, 101.

26. Kenneth N. Waltz, *Man, the State, and War: A Theoretical Analysis*, 2nd ed. (New York: Columbia University Press, 2001).

27. Ben Connable et al., *Will to Fight: Analyzing, Modeling, and Simulating the Will to Fight of Military Units* (Santa Monica, CA: RAND Corporation, 2018), 12.

28. Quoted by Garry D. Brewer and Martin Shubik, *The War Game: A Critique of Military Problem Solving* (Cambridge, MA: Harvard University Press, 1979), 52.

29. Ben Buchanan, *The Cybersecurity Dilemma: Hacking, Trust and Fear Between States* (Oxford: Oxford University Press, 2017).

30. Perla, *Art*, 4.

31. Bernard de Koven, *The Well-Played Game* (Cambridge, MA: MIT Press, 2013), 41–42, 71–73.

32. Ellen J. Langer, *The Power of Mindful Learning* (Cambridge: Perseus Books, 1997), cited by Marc Prensky, "The Motivation of Gameplay: The Real Twenty-First Century Revolution," *On the Horizon* 10 (2009), 9.

33. Dan Rea, Kelly Price Millican, and Sandy White Watson, "The Serious Benefits of Fun in the Classroom," *Middle School Journal* 31 (2000): 24.

34. Bob Stremba and Christian A. Bisson, "Teaching Theory, Facts, and Abstract Concepts Effectively," in *Teaching Adventure Education Theory: Best Practice*, edited by Bob Stremba and Christian A. Bisson (Champaign, IL: Human Kinetics, 2009), 19.

35. Nina Iten and Dominik Petko, "Learning with Serious Games: Is Fun Playing the Game a Predictor of Learning Success?" *British Journal of Educational Technology* 47 (2016): 162.

36. Johannes Breuer and Gary Bente, "Why So Serious? On the Relation of Serious Games and Learning," *Journal for Computer Game Culture* 4 (2010): 12.

37. Jerry Palmer, *Taking Humour Seriously* (London: Routledge, 1994), 3.

38. Palmer, *Taking Humour Seriously*.

39. Simon Critchley, *On Humour* (London: Routledge, 2002).

40. Heather Baird and Nicky Lambert, "Enjoyable Learning: The Role of Humour, Games, and Fun Activities in Nursing and Midwifery Education," *Nurse Education Today* 30 (2010): 548–52; Phyllis Guthrie, "Knowledge Through Humour: An Original Approach for Teaching Developmental Readers," paper presented at Annual Meeting of National Institute for Staff and Organizational Development, International Conference on Teaching and Leadership Excellence, Austin, May 23–26, 1999.

41. Claire Dormann and Robert Biddle, "Humour in Game-Based Learning," *Learning, Media and Technology* 31 (2006): 416.

42. Carol Cohn, "Sex and Death in the Rational World of Defense Intellectuals," *Signs* 12 (1987).

43. Perla, *Art*; Dunnigan, *Wargames*; Philip Sabin, *Simulating War: Studying Conflict through Simulation Games* (London: Continuum, 2012); Pat Harrigan and Matthew G. Kirschenbaum, eds., *Zones of Control: Perspectives on Wargaming* (Cambridge, MA: MIT Press, 2016).

44. Andreas Haggman, "Cyber Wargaming: Finding, Designing, and Playing Wargames for Cyber Security Education" (doctoral thesis, Royal Holloway, University of London, 2019).

45. HM Government, "The UK Cyber Security Strategy: Protecting and Promoting the UK in a Digital World," 2011.

46. National Cyber Security Centre, "Cyber Security: Fixing the Present So We Can Worry About the Future," address by Ciaran Martin, November 15, 2017, www.ncsc.gov.uk/news/cyber-security-fixing-present-so-we-can-worry-about-future.

47. Information Security Group, "Visualisation of Security via Gameplay," *YouTube*, www.youtube.com/watch?v=sPwZKB8ZLyM.

48. Marcus Carter, Mitchell Harrop, and Martin Gibbs, "The Roll of the Dice in Warhammer 40,000," *Transactions of the Digital Games Research Association* 1 (2014): 18–21.

49. June Cotte and Kathryn A. Latour, "Blackjack in the Kitchen: Understanding Online Versus Casino Gambling," *Journal of Consumer Research* 35 (2009): 742–58.

50. Patrick Zabalbeascoa Terrán, "Developing Translation Studies to Better Account for Audiovisual Texts and Other Forms of New Text Production: With Special Attention to the TV3 version of *Yes, Minister*" (doctoral thesis, Universitat de Lleida, 1993), 322–25.

51. Klaus Dodds and Philip Kirby, "It's Not a Laughing Matter: Critical Geopolitics, Humour and Unlaughter," *Geopolitics* 66 (2013): 56.

52. Cohn, "Sex and Death."

53. Matthew Kotzen, "The Normativity of Humour," *Philosophical Issues* 25 (2015): 408.

9

THE *CYBER 9/12 STRATEGY CHALLENGE*: FEEDBACK LOOPS FOR STUDENT SUCCESS

SAFA SHAHWAN EDWARDS AND FRANK L. SMITH III

Christian Van de Werken graduated from the University of Colorado with a degree in finance. Since he was also interested in law, Van de Werken took a job with the Justice Department, where he learned about organized crime and cryptocurrencies. He then pursued graduate studies in international and public affairs at Columbia University. He was mentored by Jason Healey, founder of the Atlantic Council's Cyber Statecraft Initiative and creator of the *Cyber 9/12 Strategy Challenge*—a wargame competition about cybersecurity. With Healey's guidance, Van de Werken organized the first Cyber 9/12 competition in New York and competed in the national Cyber 9/12 in Washington.

What stands out from this student's journey with Cyber 9/12? Learning from the wargame scenario was an important part of Van de Werken's experience. But even more valuable was the professional feedback that he received from the practitioners and policymakers who judged the competition. Leveraging his experience with Cyber 9/12 and expertise in cryptocurrencies, Van de Werken was soon hired by IBM as a blockchain manager. Closing the feedback loop and giving back, he has returned to Cyber 9/12 as a judge, helping adjudicate the wargame and guide the next generation of students.

Van de Werken's story is not unique. If you are a student considering this event, you are not alone. More than 5,000 students have participated in Cyber 9/12 over the years in the United States, Europe, Africa, and Australia. Many report learning similar lessons as Van de Werken, finding similar inspiration, and similarly using this experience to help land a desirable job. Moreover, many of these students did not start with traditional degrees in computer science or information technology. From finance and law to arts and social sciences, they nevertheless have learned how to connect their education and interests with technical expertise about cyberspace.

These success stories have important implications for student players, as well as for government, industry, and academia. Government and industry employers lament the so-called cyber talent pipeline problem. The talent available is scarce. Demand far outstrips supply. More than 500,000 cybersecurity jobs are currently unfilled in the United States.[1] In 2021, the global cyber workforce shortage was estimated to exceed 2.7 million unfilled positions.[2] Increasing education in science, technology, engineering, and mathematics can help increase supply over time. Yet this often-touted solution is mismatched with immediate needs, both for talent and for bridging the divide between people who develop new technology and those who use and regulate it. Attracting people with more diverse backgrounds into cybersecurity can open vast and untapped pools of talent. Interdisciplinary events like Cyber 9/12 not only prove that this is possible; they also provide employers with a mechanism for building a larger and more capable cyber workforce.

For their part, educators lament the difficulty of teaching cybersecurity due to the complex and technical content. Speaking from personal experience, we can attest to these challenges. However, cyber wargames can help teachers—and their students—connect seemingly sterile facts and abstract concepts to the real world. Like other chapters in this book, the story of Cyber 9/12 illustrates how to create safe spaces to learn about hard problems, distill complex information, formulate creative solutions, and communicate new ideas. In addition, this chapter provides insight into the generative and inspirational effects of feedback loops in educational gameplay. Feedback can help students succeed inside the classroom and beyond, building bridges that cross over into careers they may not have previously imagined. As a result, we argue that these feedback loops are important beyond their pedagogical value. They are also of practical importance for students who want to increase their professional opportunities.

THE ORIGIN STORY

In 2012, Jason Healey was in Washington, directing the Atlantic Council's Cyber Statecraft Initiative. He saw many "capture-the-flag" competitions and cybersecurity hackathons for students majoring in computer science and information technology. These events represented one way to build the cyber talent pipeline. However, they focused almost exclusively on the "how" of cyber tactics, lacking consideration of the "why," namely, the social, political, and economic context. Cryptographic puzzles, web exploits, and other technical feats are important parts of cybersecurity. But they are not the whole endeavor. Nor do they exist in a vacuum. Healey wondered if capture-the-flag competitions were sacrificing too much context in favor of an overly technical focus. That focus might be unrealistic. It might also be an unnecessarily high barrier to entry for students who might otherwise be interested in cybersecurity.

Healey asked, what if there were a cyber wargame to teach students about the complexity of cyber crisis and conflict? What if, by addressing multifaceted issues such as national security, law, and business, an *interdisciplinary* wargame could help a more diverse and inclusive array of students learn about cybersecurity? If so, the students who participated could benefit directly. Moreover, their knowledge would help develop the talent pipeline of technically literate professionals, ideally, closing the cyber skills gap as well as gaps between good technology and effective public policy.

Cyber 9/12 was his answer. Granted, the first version of this event was very different from the competition today. Healey had a small budget, and the initial format resembled a master class or table-top exercise more than a full blown cyber wargame per se. Students and faculty watched a handful of experts walk through a fictional scenario in which Iran damaged undersea cables and caused a digital blackout. The experts explained how they would respond in the various roles they might play in real life.

On one hand, this format was educational. It offered the audience insight into decision-making by senior policymakers. On the other hand, the audience was more of a passive observer, "watching the pros do it," and less of an active participant. In the years that followed, Healey and his colleagues flipped the format, placing students in the role of decision-makers and explicitly engaging them in crafting policy responses to crisis scenarios.

Though modest in origin and fluid in design, this event offered something new. Other kinds of educational simulations were well established (e.g., Model United Nations), and the military had wargamed cyberattacks for more than a decade (e.g., Eligible Receiver in 1997).[3] Yet rarely had students at civilian colleges and universities had the chance to grapple with complex cyber crises and receive professional feedback. The students who participated in the first Cyber 9/12 responded positively, indicating promise beyond a one-off event. Stakeholders in government, industry, and academia also saw value in this kind of competition. To be blunt, money talks: forward-thinking sponsors helped fund the growth of Cyber 9/12 and, critically, keep participation free for the students.

Cyber 9/12 quickly grew from one annual event in Washington into a series of competitions around the world. In 2014, the first international competition was held in Switzerland, in partnership with the Geneva Centre for Security Policy. This expansion overseas sought to increase the accessibility of interdisciplinary cyber education. It also helped the wargame address other geopolitical settings. For instance, the first competition in Switzerland addressed a cyber crisis for the European Union and NATO.

Interestingly, competitions outside the United States continued to attract American teams, particularly from US military academies. The US Military Academy at West Point, the US Naval Academy, and the Air Force Cyber College have all sent student teams overseas to gain international experience, learn about different policy frameworks, and practice multinational crisis management. For example, in 2017, the first *Indo-Pacific Cyber 9/12 Strategy Challenge* was held in Australia, hosted by the Sydney Cyber Security Network. While their Australian counterparts were not far behind, an American team from West Point managed to win the inaugural competition Down Under. In 2018 the competition expanded to the United Kingdom, and in 2019, to France. In each instance, these events demonstrated the appeal of interdisciplinary cyber wargaming for military and civilian students alike.

GAME OBJECTIVES AND PLAYERS

"I wish Cyber 9/12 was an option when I was in undergrad or law school," explains Camille Stewart, currently a deputy in the Office of the National Cyber Director, formerly the global head of product security strategy at Google, and an expert judge for this competition. "Policy focused cyber exercises were unheard of at the time. But this kind of training and glimpse into the complexity of responding to significant cyber incidents is exactly the training future cyber leaders need."[4]

At its core, Cyber 9/12 is an educational wargame. The primary objectives are to help teach undergraduate and graduate students critical thinking, problem solving with public policy, resource management, and effective communication. Each competition requires students to analyze a different cyber crisis. As a player, you must think through multiple aspects of the crisis (e.g., its technical, social, political, legal, ethical, and economic ramifications), as well as respond in limited time using realistic policy instruments. Said another way, you cannot simply admire the problem. Players are expected to propose viable solutions.

Players are also expected to effectively communicate their solutions, both with their teammates and to expert judges like Stewart and Van de Werken. Improving students' communication skill is a critical objective. Communication connects the game experience and real life, where the ability to clearly convey complex information about cybersecurity is in particularly short supply. Unlike many real-life situations, however, Cyber 9/12 gives students a safe space to make mistakes and learn. Judges are also expected to communicate, both through verbal feedback to the students and through their written scorecards, evaluating each team's presentation and policy recommendations.

Yes, winning is a desirable goal for most players. But it is not necessary for learning. Losing teams stand to learn as much, if not more. And winning is not emphasized over participation in the competition.[5] Players are encouraged to leverage this competition to consider creative ideas and voice innovative solutions. Bold moves are welcome. Even when they do not work, novel policy proposals can still provide teachable moments. Surprisingly, students' creativity has proven useful for some judges as well. Judging Cyber 9/12 has helped some subject matter experts shake up their own assumptions in a friendly and low risk environment. Some judges have collected briefing materials from student teams to reference potential solutions from the game to the real challenges they grapple with in their daily job. In recent years, the Atlantic Council has also included illustrative examples of effective policy recommendations from prior competitions into the playbooks that students use to prepare.[6] These examples help students learn from teams in other games. They can also help inspire judges and sponsors who are on the lookout for new solutions to cyber challenges in real life.

Players, judges, and sponsors often have additional objectives beyond education. A range of other implicit and explicit motives are at work, analogous to those described by Rachael Shaffer in chapter 7. Some students are motivated by free food (e.g., ham, cheese, and truffle sandwiches were popular in Sydney), or the awesome view from the BT Tower in London. More seriously, some students are most interested in the professional coaching and contacts they can gain through the career fairs, résumé workshops, and industry receptions offered alongside this competition. Some judges are likewise interested in professional networking. This event allows them to meet new people or reconnect in a more congenial setting than the typical conference or trade show. For their part, government and industry sponsors are typically interested in recruitment. Because Cyber 9/12 provides special access to a diverse talent pool, sponsors have offered job interviews to competitors and, in some cases, tentative job offers during the event. Fortunately, these additional objectives usually complement rather than clash with the primary educational objectives of this wargame.

REALISTIC FAKES? SCENARIO DESIGN, GAME SEQUENCE, AND ROLE PLAY

Every Cyber 9/12 scenario is different. Each is designed from the ground up by the Atlantic Council and the competition partner, making this a living game series that continues to evolve. Given the educational objectives, scenario design emphasizes player engagement and realism over analytic utility. Little systematic attempt has been made to collect research data on the policies that student teams have proposed over the years, or on how judges adjudicate these proposals from game to game. Instead, Cyber 9/12 scenarios focus on engaging or immersing students in a fictional crisis wherein they play the role of senior policy advisers to a national or supernational government. These scenarios incorporate pertinent themes and cyber policy challenges for which local stakeholders in government and industry need creative solutions.

On one hand, engagement and realism can go hand in hand, at least to the extent that students and judges are most interested in scenarios that they can easily imagine happening in real life. Scenario elements that mirror events drawn from headline news—such as a ransomware attack on European hospitals or a software supply chain attack on maritime shipping—can inspire and excite players by putting them in the shoes of the policymakers that they read about every day in real life.[7] Formulating and showcasing the policies they would propose in realistic scenarios helps players learn (judges, too).

On the other hand, realism is limited to the extent that some technical and social aspects of the scenario are abstracted or simplified. Simplification helps keep the competition tractable in the limited time available. It also helps keep players with different backgrounds engaged enough to learn. Too much technical detail, and students without technical backgrounds may disengage; too much social detail, and students who lack exposure to political science or related fields may disengage. Of course, if there is too little detail on any of these dimensions, then students who have relevant social or technical expertise may "fight the scenario" rather than suspend their disbelief and play the game.[8] Scenario design is therefore a balancing act between deliberately challenging—perhaps even overwhelming—students with different backgrounds enough to learn outside their comfort zone while not frustrating them so much that they give up or nitpick.

The balance of detail provided also varies over the course of each competition. In Cyber 9/12, students respond to a three-part scenario. A month before the competition, they receive their first "intelligence report." This report consists of about twenty pages of background material, setting the scene and describing the initial phase of the emerging crisis. Because students have several weeks to digest it, this report includes detailed technical information. This helps set high expectations—the competition is not easy.

Successful players use their lead time to conduct independent research and consult with their coaches on the social and technical issues raised in the initial report. These issues might include forensic malware analysis, for instance, or the design for supervisory control and data acquisition systems used in a power plant, coupled with the politics surrounding freedom of navigation operations in the South China Sea, or the division of labor between local and federal law enforcement. The information provided in the intelligence report is deliberately incomplete. It also includes red herrings. These elements help teach students about the uncertainties and misleading information that are hallmarks of real-world policymaking about cybersecurity.

The interpretation of cyberspace in Cyber 9/12 is partially a function of discourse and debate within each team, as well as between the teams and the judges. In other words, it emerges as part of gameplay. For example, an intelligence report may provide a detailed schematic of the information technology in a car, targeting data for ransomware, narrative description of a white hat hacker organization, or the command line code for a malicious Linux exploit. Sensemaking or interpreting how these related but distinct details about cyberspace interact with other sociotechnical systems is part of the challenge. Decision-making is neither rigid nor preprogrammed. This free or open aspect of gameplay allows for a broad interpretation of what "cyber" is.

That said, the scenario is preset, and the scenario design tends to highlight the interconnected aspects of cyberspace. Competitors and judges do not know what will happen next until they receive the second and third intelligence reports. The content of these injects does not change based on the students' policy proposals. The sequence of events is outside their control, and the range of interconnected policy challenges grows over the course of competition. Students therefore learn to update their analysis and recommendations based on the scenario and judges' feedback over successive rounds of play.

Each competition consists of three rounds. These rounds roughly correspond to the three core tasks of reporting, responding, and reacting to the scenario. In the initial qualifying round, student teams submit a written policy brief based on the first intelligence report. Then, during the qualifying round of the in-person or virtual competition, they deliver an oral briefing about their policy proposals before a panel of judges. Teams must leverage what they learned in the previous month; they must also think quickly on their feet to answer questions from the judges. Learning objectives for this round of competition include analyzing, synthesizing, and reporting cyber incidents to senior officials.

After the qualifying round of competition, everyone receives a second intelligence report on the unfolding crisis. Those teams that advance to the semifinal round have less than 24 hours to prepare their responses. They learn how to adapt their background research and initial recommendations based on new information. Agility is advantageous. At this stage, teams are expected to update their policy proposals and then communicate increasingly detailed recommendations to the judges. Fifteen minutes before the final round of competition, a third, shorter intelligence report is provided. The finalists must, in turn, react quickly to this new information and, under high pressure, demonstrate their mastery of strategic thinking to the judging panel.

A Cyber 9/12 scenario usually starts with a slow burn. For instance, for the 2020 competition in Austin, the first intelligence report described a software vulnerability that was discovered by a white hat hacker and disclosed to the vendor.[9] At this stage, student teams might recommend policies to slow exploitation of this vulnerability. In the second intelligence report, the vendor prepared to patch the vulnerability in the next software update, scheduled for the following week. However, a malicious actor was already exploiting the vulnerability to manipulate the power grid and cause blackouts on election day. At this point, teams must quickly respond to an increasingly serious technical, business, and political crisis. The final intelligence report featured more international connections. Manipulation of the power grid was tied to a foreign government, and students must devise diplomatic and military options in response.

Along with the interconnected aspects of cyberspace, another common feature is crisis. Crisis is inherent to the name of Cyber 9/12—harking back to the day after the terrorist attacks on September 11, 2001. Admittedly, the cyber threat to date does not resemble a "cyber 9/11" or a "cyber Pearl Harbor."[10] Audiences outside the United States are also less likely to recognize the reference to 9/11. More than two decades later, younger Americans may not immediately make the connection either. That said, Cyber 9/12 has become a recognizable brand in its own right. As part of the trade-off between player engagement and realism, scenario design also balances between conveying a sense of crisis or urgency to excite players and elicit action while at the same time avoiding threat inflation.

Most scenarios include or implicate intentional and malicious action. Whether state or nonstate, profit-minded or politically motivated, fictional antagonists leverage cyberspace to help achieve their goals. As in real life, however, attribution is not always clear. Nor are offensive cyber operations always one-sided. For instance, in the 2017 Indo-Pacific competition, student teams provided policy advice to the Australian government and private sector during a crisis in which Australia and its allies were implicated in hacking back against China. During the last round of this competition, finalists also had to respond to a pipeline explosion while the cause of the blast—accident or attack—was still unclear. The strategic ambiguity evoked by these kinds of scenario elements helps students think critically about the technical and political challenges of attribution, along with the pros and cons of various responses to different kinds of malicious cyber activity and actors. Managing uncertainty is a core part of the wargame.

The policy responses that students consider include many of the same options that the competition judges grapple with as decision-makers and subject matter experts in real life. In Cyber 9/12, judges play the role of senior policymakers, usually in government. This is not a stretch. Past judges have included real politicians, generals, ambassadors, chief executive officers, analysts, professors, and intelligence officials, among others. Depending on where the competition is held, the panels of judges nominally represent the Principals Committee of the National Security Council (United States), the Secretaries Committee (Australia), the Political and Security Committee (European Union), or the Civil Contingencies Committee (United Kingdom). Playing one of these roles, the judges listen to players' presentations and ask follow-up questions. Student teams play the role of senior government advisers. In the United States, this usually means staff on the White House's National Security Council. In Australia, student teams take on the roughly analogous role of cyber policy experts within the Department of the Prime Minister and Cabinet. In Geneva, students play the role of senior cybersecurity experts invited to brief the EU Political and Security Committee. Students in the United Kingdom act as policy advisers in the Prime Minister's Office.

After each team's policy presentation and a few minutes of questions and answers, both judges and students step out of their assigned roles for an open conversation. This allows the judges to provide constructive feedback. By "breaking the fourth wall" after role play and allowing for more two-way communication, students get personalized advice and professional guidance from senior practitioners and policymakers. Students often describe this as a heady and inspiring experience. Judges likewise report their experience to be overwhelmingly positive. The educational objective and fictional nature of these competitions allow even current government officials to let their hair down and speak with students—as well as their fellow judges—with greater ease and authenticity than their everyday jobs typically allow. The feedback loop between players and judges is crucial for students to learn, grow, and improve their understanding of cybersecurity.

ADJUDICATION AND VICTORY CONDITIONS

Cyber 9/12 competitions are adjudicated by subject matter experts on the judging panels. For each team presentation, each judge fills out a scorecard. Points are awarded based on the teams' understanding of cyber policy, identification of key issues, analysis of policy responses, communication skills, and creativity. Given the intelligence reports and their own expertise, the judges on each panel decide amongst themselves about the feasibility and efficacy of the cyber and noncyber actions proposed by the student teams. There is no formal model or combat simulation running in the background. For better or worse, success or failure is a judgment call, informed by the expertise and experience of each panel of judges.

The scores provided by individual judges are averaged and compared, with the highest-scoring teams advancing to the next round of competition. Judges also nominate teams for recognition in categories such as "best teamwork" and "best policy briefing." Again, while winning is important, it is not the only objective of the game. Teams that do not advance to the semifinal and final rounds still receive the second and third intelligence reports, and they are still encouraged to watch and learn from the remaining competition.

What distinguishes the teams that win? According to most judges, the best players demonstrate comprehensive knowledge about the crisis; rigorously analyze the potential implications; effectively communicate both technical and nontechnical content; and provide clear—ideally creative and actionable—policy recommendations. Teams composed of players with multidisciplinary skillsets tend to perform well. Coaching also helps. Every team is strongly encouraged to recruit a coach or mentor to help them prepare for competition. Many coaches are teachers at the students' college or university; others are practitioners from government or industry. Like judges, all coaches volunteer their time. As mentors, they can make a significant difference in how well their team performs under the pressure of competition. Coaches also serve as another valuable source of professional feedback and advice, as well as continuity, increasing the likelihood of teams returning to compete again.

Winning is hard work. For instance, several teams from West Point—named the Black Knights—have performed well over the years, gaining a reputation for success, both in the United States and overseas. These teams do well in part because of their coaching and mentorship. They are also familiar with dedicated teamwork and blunt feedback. Successful teams do their homework on the relevant roles, responsibilities, and capabilities of government agencies, so they are prepared to direct their policy recommendations to the appropriate organizations. None of these attributes are unique to military cadets. Well-prepared students from civilian schools are equally competitive.

As the competition has grown, it has attracted graduate students who are also midcareer professionals. Some of these students have prior experience in national security, delivering executive briefings, and even responding to cyber crises inside government and industry. To help foster a culture of inclusion and fair play, the Atlantic Council now splits the competition into a student track and a professional track. Teams that include someone with more than two years of prior work experience in cybersecurity or public policy compete in the professional track. This change was made in 2019 for the competition in Washington, which tends to attract the most midcareer professionals. That way, a freshman from Brown University will not compete against a graduate student from the National Defense University who already has years of experience working at the Pentagon. About a quarter of teams that compete in Washington are qualified for the professional track. The two-track model still encourages more and less experienced players to interact and learn from each other over the course of the competition.

CHALLENGES, MISTAKES, AND LESSONS LEARNED

Cybersecurity is about more than hardware, software, and data. So is educating and recruiting the next generation of innovators and leaders or, from the student's perspective, becoming an innovator and leader yourself. Cyber 9/12 helps address these complex challenges by bridging several important gaps: social and technical, academic and professional, military and civilian, theory and practice.

Bridging these gaps is not easy, and there have been missteps and missed opportunities along the way. For example, universities often pay lip service to interdisciplinary education, as well as diversity and inclusion in tech. But talk is cheap. Support for concrete action is harder to come by, and funding for cyber wargames remains relatively rare in civilian higher education.[11] Sometimes, other stakeholders in government and industry also have a hard time deciding what to make of Cyber 9/12. Even when potential sponsors see the value proposition, some have trouble providing financial support for this kind of interdisciplinary event. Innovation is hard—and innovative sponsorship, donations, and grants are no exception.

Fortunately, Cyber 9/12 has developed a community of committed stakeholders who are familiar with this event and the value it provides for students, educators, and employers. They include financial sponsors, as well as hundreds of judges and coaches who have lent their time and expertise over the years to help educate and inspire students through these cyber wargames. Most important, this community includes the student players, who take the initiative to step up and learn through this competition.

While Cyber 9/12 can continue to grow, sustainability is a challenge on at least two fronts. The first is variation in the background knowledge that different players bring to the game. Some players are novices. They come to compete with little direct experience. But others are cybersecurity practitioners and professionals in public policy. This variation is good. It means that these competitions attract newbies and professionals alike and that, at the professional end of the spectrum, more people now have more experience and expertise in this critical field. Nevertheless, when it comes to student preparation and fair competition, this variation is also a challenge.

Separate tracks of competition are one way to help level the playing field. In addition, in order to improve learning and give less experienced teams a fair shot, the Atlantic Council now provides more information in advance through playbooks that review previous games and policy proposals from award-winning teams.[12] This information also includes resources tailored for novice teams, such as templates for their outreach to potential coaches. All players are well advised to take full advantage of these resources. Still, as noted in chapter 1, variation in background knowledge and experience remains a challenge for fair play. Judges vary as well. The best judges calibrate their feedback to educate and inspire students with different levels of experience so as to neither underwhelm nor overwhelm them.

A second challenge is scale. The demand for cyber talent continues to grow, which has resulted in calls for more Cyber 9/12 competitions in more locations. Here again, high demand is a good sign and testament to the value of these wargames. Being able to meet this demand while preserving quality control is a challenge. Recall that each wargame involves a bespoke scenario, tailored to a particular time and place. Each competition also requires its own advertising, recruitment, communication, coordination, fundraising, organization, and execution. Therefore, sustainment and expansion depend on considerable support from local partners. Capable partners cannot be taken for granted.

By addressing national security and public policy, Cyber 9/12 was originally envisioned as an interdisciplinary alternative to narrowly technical hackathons and capture-the-flag competitions. A decade later, it boasts alumni across government, academia, and industry. Many have backgrounds other than computer science or information technology. Yet they now do important work in cybersecurity. Not only are students gaining the skills and confidence to transition from a degree in the humanities to work at the US Cyber Command, for example, or a cybersecurity firm. Employers in government and industry are also learning that technical degrees are not always necessary for employees to excel in this field. More students with technical backgrounds are competing in Cyber 9/12 as well, seeking to hone their policy analysis and communications skills. Of course, more work needs to be done to grow the supply of cyber talent; to bridge divisions between social and technical expertise; and, if you are a student reading this, to build your knowledge base and professional opportunities. Hopefully, Cyber 9/12 can help for years to come.

NOTES

1. "Cybersecurity Supply/Demand Heat Map," Cyber Seek, no date, www.cyberseek .org/heatmap.html.
2. National Initiative for Cybersecurity Education, "Cybersecurity Workforce Demand," www.nist.gov/system/files/documents/2021/12/03/NICE%20FactSheet_Workforce %20Demand_Final_20211202.pdf.
3. On Eligible Receiver, see Jason Healey and Karl Grindal, eds., *A Fierce Domain: Conflict in Cyberspace, 1986 to 2012* (Vienna, VA: Cyber Conflict Studies Association, 2013), 40; and Fred Kaplan, *Dark Territory: The Secret History of Cyber War* (New York: Simon & Schuster, 2016), chap. 4.
4. Correspondence with the author, May 2021.
5. The relationship between competition and learning is complicated. Among others, see Valentin Riemer and Claudia Schrader, "Playing to Learn or to Win? The Role of Students' Competition Preference on Self-Monitoring and Learning Outcome When Learning with a Serious Game," *Interactive Learning Environments* 30, no. 10 (2022), 1838–50; and Nergiz Ercil Cagiltay, Erol Ozcelik, and Nese Sahin Ozcelik, "The Effect of Competition on Learning in Games," *Computers and Education* 87 (2015). Competing to win can also be distinguished from competing to excel, and different types of competitiveness may resonate with different students. E.g., see David R. Hibbard and Duane Buhrmester, "Competitiveness, Gender, and Adjustment among Adolescents," *Sex Roles* 63, no. 5 (2010).
6. Atlantic Council, *Austin, Texas Cyber 9/12 Strategy Challenge Playbook*, 2020, www .atlanticcouncil.org/content-series/cyber-9-12-project/austin-texas-cyber-9-12 -strategy-challenge-playbook/.

7. E.g., see Nicole Perlroth and Adam Satariano, "Irish Hospitals Are Latest to Be Hit by Ransomware Attacks," *New York Times*, May 20, 2021; and Ellen Nakashima, "Russian Military Was Behind 'NotPetya' Cyberattack in Ukraine, CIA Concludes," *Washington Post*, January 12, 2018.

8. On the willful suspension of disbelief, see Peter P. Perla and E. D. McGrady, "Why Wargaming Works," *Naval War College Review* 64, no. 3 (2011): 113.

9. Atlantic Council, *Austin, Texas Cyber 9/12 Strategy*.

10. For a critical analysis of catastrophic cyber scenarios, see Myriam Dunn Cavelty, *Cyber Security and Threat Politics: US Efforts to Secure the Information Age* (New York: Routledge, 2008); and Sean T. Lawson, *Cybersecurity Discourse in the United States: Cyber-Doom Rhetoric and Beyond* (New York: Routledge, 2019).

11. The same can be said about wargames in general. See Philip Sabin, "Wargaming in Higher Education: Contributions and Challenges," *Arts and Humanities in Higher Education* 14, no. 4 (2015): 329–48.

12. Atlantic Council, *Austin, Texas Cyber 9/12 Strategy*.

10

THE EVOLUTION OF THE NORTH AMERICAN ELECTRIC RELIABILITY CORPORATION'S GRID SECURITY EXERCISE

MATTHEW DUNCAN

Electricity is arguably the most "critical" of critical infrastructure. It is essential to modern economic and social well-being; moreover, most other kinds of critical infrastructure also depend on electricity. These dependencies make the North American bulk power system (BPS) among the continent's most important physical and cyber systems. The countries' economies and nearly 400 million people in the United States and Canada rely on the thousands of power generation and transmission assets that constitute the BPS to provide electricity on demand.[1] Unfortunately, nation-state adversaries—including Russia and China—may threaten the reliable flow of electricity to achieve military and economic advantage.[2] Nonstate actors can also threaten the BPS, whether by accident or by design.

The North American Electric Reliability Corporation (NERC) and its Electricity Information Sharing and Analysis Center (E-ISAC) are delegated authority from the US federal government and Canadian provinces to oversee the reliability, security, and resilience of the BPS. One way they do so is through a biennial Grid Security Exercise (*GridEx*). *GridEx* plays an important role in cyber and physical defense training and, more generally, in the collective defense of North America, by providing utilities and government partners with the opportunity to practice their response and recovery from attacks by capable adversaries.

GridEx is the largest exercise or wargame of its kind. It takes real and fictional events and turns them into complex cyber and physical security scenarios. These scenarios are designed to stress—if not overwhelm—even well-prepared utility providers and government agencies. These challenging scenarios motivate players to innovate in order to restore the power grid, thereby testing their established plans and procedures as well as inspiring new and creative thinking. Since 2011, *GridEx* has evolved from a one-room tabletop exercise into a major event for thousands of practitioners and senior officials in government and industry.

HOW IT BEGAN

Drills to practice restoring power in the aftermath of accidents, errors, and equipment failures have long been standard practice in the electricity industry. Real-world events often inspired these exercises, including natural disasters and the large-scale power blackouts in the American Northeast in 1965 and 2003.[3] The importance of critical infrastructure protection, and the vulnerable relationship between physical and cyber-based systems, drew increasing attention during the 1990s, along with the possibility of deliberate and malicious attacks against these systems. Later, concerns about the risk of terrorism became more acute after the September 11, 2001, terrorist attacks. Since then, awareness of attacks in and through cyberspace as an effective means to disrupt the power grid grew.[4]

These vulnerabilities and threats are of particular interest for NERC. Focused on energy reliability, NERC develops and enforces reliability standards. These include critical infrastructure protection standards that provide a baseline of cyber and physical hygiene, which electricity utilities must in turn comport. After revelations in 2010 about the Stuxnet worm (i.e., a cyberattack that targeted industrial control systems in Iran), serious questions were raised about the cybersecurity of the electricity grid in North America.[5] Similar concerns remain prevalent today. For example, in April 2022 it was reported that nation-state adversaries were developing the cyber equivalent of a "Swiss Army knife" to attack industrial control systems used by the electricity industry.[6]

To help prepare an effective response, NERC and its E-ISAC launched *GridEx* in 2011. The objective was to strengthen industry and government's response to a cyber incident. Initially, *GridEx* was a tabletop exercise. The exercise consisted of a continent-wide scenario, with seventy-five industry and government organizations from the United States and Canada participating. The scenario featured a fictional cyberattack by an advanced persistent threat actor. The attacker was able to inject malicious code (malware) into the BPS by compromising the industrial control systems used by many utilities. The utilities that were unable to detect and mitigate the intrusion experienced intermittent power outages that had an impact on millions of people.[7] In response to this scenario, operators from the major functions of electric generation, transmission, and distribution discussed how to put the BPS back together after such an attack.

The initial *GridEx* highlighted industry's existing individual cyber incident response plans, and industry-wide cooperation. However, it was discovered that, in many instances, communications across teams, other critical infrastructure sectors, the supply chain, and government were limited—not integrated into their procedures. Many participants did not know who to call, or which federal or state agency had the capabilities or jurisdiction to assist. Furthermore, industry had not made sufficient investments in information sharing, limiting the value that E-ISAC and other sharing platforms could offer the grid during a significant cyberattack. Participants saw value in the tabletop exercise, but they also recognized the need to broaden and deepen their discussions from the policy level to real response processes and procedures at their utilities. Talking around the table could only improve cyber response so much.

While somewhat troubling, the lessons learned through this initial exercise were instructive and inspired the creation of what has since become the *GridEx* format of distributed play and executive tabletop. Moreover, by identifying potential problems and training for the response, *GridEx* demonstrated the value of cyber wargaming for the electricity industry to strengthen relationships, enhance coordination, and improve communications across sectors and government. *GridEx* was institutionalized as a biennial event as a result.

In 2013, NERC and the E-ISAC expanded the scope of *GridEx II* into the format used today. The second iteration incorporated additional stakeholders from outside the electricity industry, such as natural gas, telecommunications, water, and finance. It also began offering a "distributed play" option, giving participating utilities their own free tabletop exercise in a box.[8] Since then, the objectives for *GridEx* have included exercising emergency response plans; increasing participation of local and regional response organizations; improving coordination among interdependent infrastructures; increasing supply chain participation; improving communication; engaging senior industry and government leaders; and gathering lessons learned. This expanded participation and scope not only improved industry resilience but also presaged the rapidly evolving and increasingly dangerous threat environment.

EXERCISE SCENARIOS IMITATE LIFE, WHICH IMITATES EXERCISES

Each iteration of *GridEx* features a different scenario, developed in concert with industry and government planners. These scenarios are intended to reflect increasing threats to the North American electricity grid. While the first *GridEx* was conceived primarily as a cyber wargame, it soon became apparent that including both cyber and physical threats was the best way to train utility operators to respond to complex attacks. Said another way, *GridEx* evolved from a wargame *about* cyber to a broader wargame *with* cyber.

For example, during the real-world Metcalf substation incident in April 2013, one or more snipers shot at seventeen high-voltage transformers in Coyote, California.[9] These rifle shots caused little harm to the substation's electricity supply, and they had no impact on the overall reliability of the BPS. However, this attack still resulted in more than $15 million in damage. The attack's apparent complexity, coordination, and precision also raised concerns within industry and government. While the attacker(s) was never identified, this incident demonstrated that electricity infrastructure in North America was vulnerable to physical attack. Coincidently, in 2013, planners working on the *GridEx II* scenario included a combined physical and cyber event when the Metcalf sniper attack occurred. The incident further emphasized the need to include physical threats during the exercise. Physical attacks have continued since Metcalf, including in December 2022, when gunfire damaged a distribution level transformer and caused outages to 45,000 customers in North Carolina.[10]

Similarly, the *GridEx III* scenario, held in November 2015, involved an attack against the power supply by a fictional nation-state. One month later, Ukraine's electricity industry suffered a real-life cyberattack that affected thirty distribution-level electrical substations. The resulting blackout affected 225,000 people for several hours around Kyiv. The US government later attributed this cyberattack to Russian hackers—a group dubbed Sandworm—using malware called BlackEnergy.[11] In addition to having just recently gamed out attacks by this kind of advanced persistent threat, *GridEx* also helped foster relationships that proved useful in the aftermath of BlackEnergy and subsequent cyberattacks against the Ukrainian power grid.

In December 2016, Sandworm hackers struck again with another cyberattack against Ukraine's critical electric infrastructure. This attack focused on the transmission system. It used malware called Crashoverride/Industroyer, specifically targeting four industrial control systems used to open and close circuit breakers at high-voltage substations.[12] Russian-linked adversaries not only demonstrated their capability but also their intent to attack civilian electric infrastructure. Recognizing the growing threat to industry, the E-ISAC and SANS Institute collaborated on a public case study of the 2016 attack to better inform industry defenders of the growing threat to industrial control systems.[13]

Other real-world examples of cyberattacks that impact critical infrastructure continue to emerge. They range from a cyberattack on safety systems at a Saudi Arabian Petrochemical Facility in 2017 (Triton/Trisis), an attack on the Ukrainian financial services sector that "escaped the lab" to impact global shipping in 2017 (NotPetya), and a ransomware attack that shut down the Colonial Pipeline for gas distribution in the United States in 2021.[14] Finally, while most cyber and physical attacks by Russia around the invasion of Ukraine in 2022 took place in Eastern Europe, malware nevertheless targeted telecommunications and industrial control systems (e.g., Industroyer2), which bore some resemblance to the *GridEx VI* scenario in November 2021.[15]

To the extent that *GridEx* scenarios have anticipated some aspects of real-world events, some commentators in government and industry have referred to the darkest of potentially dark days as a "*GridEx*-level event." Of course, the goal of this exercise is training to mitigate these events, if not prevent a *"GridEx*-level event" from ever even occurring.

A CONTINENT-WIDE EXERCISE

Large exercises and wargames are challenging to plan and run. Exercising critical infrastructure such as the BPS—which is continent-wide, with multiple jurisdictions and cyber-physical systems that have different operating postures, procedures, and regulations—is particularly daunting. Preparing for potential attacks requires planners to contend with the full scope of the problem, not to mention the full cast of characters involved in the response.

Consequently, the planning process for *GridEx* involves a wide range of stakeholders. They collaborate to develop the exercise scenario and recruit representatives from industry and government. The design cycle involves participants from the Grid Exercise Working Group, which represents the E-ISAC, NERC, industry, and government. Participation of the US Department of Energy (DOE) laboratories has also helped advance distributed gameplay through *GridEx*, providing realistic training malware modeled on real-world attacks to help defenders practice recognizing and analyzing malicious code in a safe environment. The E-ISAC also provides simulated traditional and social media, including mock "newscasts" and social media posts to help utility and government public affairs personnel practice responding to media requests, misinformation, and disinformation.

As *GridEx* grew, however, so did the challenges in creating a coherent scenario for a continent-wide exercise. In 2017, for example, the *GridEx IV* distributed play scenario was decentralized, enabling all electricity organizations in North America to customize their exercise to meet individual training objectives. Each organization decided how it would participate. Starting with a baseline scenario provided by NERC, each organization appointed lead planners to help customize the scenario for their players. They also identified individual players from their organization's various business functions and coordinated activities with the other internal stakeholders. In fact, exercising interoperable elements of a utility (e.g., security, information technology, public affairs, and operations) was a critical element of the exercise, given the way utilities operate.[16] However, the decentralized nature of play in an inherently interconnected system caused confusion for neighboring utilities running "different" scenarios, limiting their and government's ability to render mutual assistance as they normally would in a real crisis.

During the exercise itself, all the various players—utilities, generation and transmission owners, reliability coordinators, independent system operators, balancing authorities, the DOE, the Department of Homeland Security (DHS), and the Federal Bureau of Investigation—receive a detailed description of the scenario via email. This description includes simulated reports (e.g., the DOE-417 or CIP-008-6), submitted via an exercise portal so that utilities can practice submitting the required information.[17] The players then use incident response measures, such as the *GridEx* Exercise Portal, to communicate within their organizations and to coordinate across the electricity sector. An exercise control cell run by the E-ISAC manages scenario distribution, monitors exercise play, and captures response activities across the virtual and distributed gameboard.

Ongoing policy conversations between senior government officials and chief executive officers on the Electricity Subsector Coordinating Council have also informed the executive tabletop portion of *GridEx*. The executive tabletop continued to focus on the policy decisions that senior industry and government leaders would make during an emergency of the scale envisioned in *GridEx*.

Over the years, for example, tabletop participants have faced important decisions about expanding the Electricity Subsector Coordinating Council's Cyber Mutual Assistance program; advancing industry collaboration with the E-ISAC; reaching unity of message among senior executives; and requesting that DOE declare a grid security emergency under the Federal Power Act, Section 215A.[18] As with the exercise as a whole, participation in the executive tabletop has steadily grown, with more than 200 senior executives from 88 industries and governments across the United States and Canada participating in a fully remote *GridEx VI* in 2021.[19]

EVOLVING OBJECTIVES AND LESSONS LEARNED

After each *GridEx*, NERC issues a public report on the lessons learned. These reports typically indicate that while most participating organizations have effective cyber incident response plans, more needs to be done to improve coordination with other stakeholders and parts of the critical supply chains. The goal is to encourage learning, continuous improvement, and participation through positive reinforcement. These reports also highlight the importance of staying current with protocols and guidelines that are responsive to evolving threats, and the need for additional training to enhance preparedness and resilience.

Similarly, the objectives for *GridEx* evolve in response to the lessons learned each year (amalgamating training, education, and policy). High-level objectives over the years have included exercising incident response plans, particularly for industrial control systems (2011–21), expanding local and regional responses (2013–21), engaging critical interdependencies (2015–21), increasing supply chain participation (2017–21), and improving communication as well as unity of message (2013–21). To illustrate, a selected sample of the key lessons from each exercise is described next.

GridEx II, *2013*

As mentioned above, the *GridEx II* scenario featured coordinated cyber and physical attacks on the BPS. The simulated cyberattack affected business and industrial control networks, similar to the real "Shamoon" cyber espionage campaign, which wiped computers' master boot records, rendering them inoperable.[20] The concurrent physical attack used improvised explosive devices and ballistic weapons to degrade the reliability of electricity infrastructure and threaten public health and safety.[21] This scenario was designed to test utilities' all-hazard crisis response and strengthen their information-sharing relationships and processes.

GridEx II emphasized the business and operational implications of isolating information technology assets during a cyber event. What lessons were learned? For one, this exercise highlighted how mechanisms to collect and preserve forensic evidence are essential after cyber and physical attacks. The exercise also indicated that emergency response plans should include clear procedures for escalating incident response recommendations to help ensure consistent responses across participating organizations. Alternative and redundant telecommunications channels are needed as well. Information provided to the public through print, radio, television, and social media could support (or hinder) government and industry efforts to restore electricity.

The executive tabletop indicated that unity of effort was essential to identify, discuss, and decide the many policy issues that arise from a severe event. Doing so requires that industry executives and senior government officials at local, state, federal, and potentially international levels are directly involved. Unlike weather events (e.g., a hurricane), which are generally known about in advance, cyberattacks present no-notice planning challenges for restoring electricity.

Similarly, the participants found that a well-coordinated physical attack presents particular challenges for how the industry restores power, such as being able to procure and safely replace damaged equipment while being shot at or bombed, or when the root cause of the failure is still unknown. Given these challenges, the electricity industry realized it needed to enhance mutual aid arrangements and critical spare equipment inventories. Finally, the executive tabletop highlighted how existing laws and statutes—including the Defense Production Act and the Stafford Disaster Relief and Emergency Assistance Act—might help the grid recover from a severe event.

GridEx III, *2015*

For the third iteration, the scenario focused on unusual and anomalous control system behavior. Cyberattacks often manifest this way and only reveal their insidious and potentially unique techniques after investigation. Players received reports from the E-ISAC of increased electric substation break-ins and unexplained drone surveillance. These initial events were exacerbated by telecommunications disruptions, which impeded the ability of generation and transmission operators to deliver electricity to their customers. A new strain of malware was discovered when equipment began behaving erratically and communications pathways were interrupted. These problems complicated the response by the incident command system, which was now unable to communicate and trust critical systems. Meanwhile, corporate communication teams were managing inaccurate social media reports and negative public attention (rival information and adversary influence operations being an innovative new addition to the exercise in 2015).[22]

This time, effective information sharing and reporting proved difficult given the redundant and time-consuming nature of some of the communication tools. Nevertheless, findings from the exercise indicated that *GridEx* scenarios should continue to prompt operations management, support staff, corporate communicators, and field operations to communicate with cyber and physical security personnel. Likewise, the exercise concluded that the E-ISAC portal and sharing capabilities needed to be improved, so that the most important information is highlighted to support effective real-time communication and collaboration.[23]

The *GridEx III* executive tabletop underscored the challenges and opportunities related to social media. In addition, industry needed to coordinate with local law enforcement to identify and assess the physical risks for electricity facilities and workers. Unlike mutual assistance in response to major storms, the industry's capability to assist in the analysis and response to malware was limited. Greater expertise was needed from software suppliers, control system vendors, and government agencies. Finally, the financial impact of a *GridEx*-level event was not lost on senior executives. If a capable nation-state launched a serious attack, utilities in North America would need unprecedented levels of financial resources to restore their facilities and eventually resume normal operations. The need for greater coordination between the electricity and financial industries was another insight from this exercise.[24]

GridEx IV, *2017*

GridEx IV saw lead planners take a more proactive role in recruiting players from their organizations, including participants from operations and communications teams, and coordinating with external organizations, such as local law enforcement, government agencies, and utility equipment vendors.[25] In 2017, the scenario posited cyberattacks against energy management systems combined with car bombs that targeted electricity generation and transmission facilities. Adding to the complexity, the adversary also attacked interdependent infrastructure for natural gas and telecommunications. These events necessitated greater coordination within and across industry, including other critical infrastructure sectors, as well as coordination with government, including the National Guard.[26]

Corporate communicators proved particularly important, especially those addressing misleading or false information in the social media sphere. The exercise reinforced the need for resilient telecommunication, with utilities prepared to use multiple technologies in the event that one or more channels of communication are lost. For example, electricity utilities can use the DHS's Wireless Priority Service and the Government Emergency Telecommunications Service.[27] A related lesson from this exercise was the recommendation that the E-ISAC become a member of the DHS Shared Resources High-Frequency Radio Program and acquire its own high-frequency radio capability.

GridEx IV was the first attempt to exercise the industry's Cyber Mutual Assistance program, developed after *GridEx III*. This program provided a pool of expert cybersecurity volunteers to support utilities in the event of a major cyber incident or attack. One lesson was that more utilities should participate in the Cyber Mutual Assistance program. Likewise, the Electricity Subsector Coordinating Council and the E-ISAC should engage more with this program to encourage both the efficient sharing of relevant information and compliance with nondisclosure agreements before, during, and after a cyber event. The Cyber Mutual Assistance program also featured in the executive tabletop as a potential avenue to share equipment, expertise, and resources when normal supply chains and service agreements are overwhelmed. Room for improvement in intergovernmental coordination between the United States and Canada was also identified.[28]

GridEx V, *2019*

Recognizing that cyber events begin long before the "lights go out," the scenario for *GridEx V* began with preexercise gameplay. This phase was simulated on an E-ISAC exercise portal where simulated surveillance and chatter about adversary attacks on European energy systems were shared. Once exercise play began, players worked to put the pieces together to guide their investigations and response to exercise malware, which replicated the CrashOverride malware used in the real 2016 Ukraine incident.[29] The cyberattack in *GridEx V* caused utilities to lose visibility on their systems, forcing them to revert to onerous and complicated manual grid operations to keep electricity flowing. At the same time, adversaries deliberately distributed disinformation on social media to instigate discord and even violence against utility workers trying to restore power. Finally, the prolonged outages caused by the attacks affected water and wastewater plants, as well as the natural gas pipeline system.[30]

GridEx V included more small- and medium-sized utilities in distributed play. These participants may have limited resources to respond independently to this type of attack. Interdependent industries—including natural gas utilities, water utilities, telecommunications companies and original equipment manufacturers—also increased their participation. All these new participants found value in the collective defense concept promoted by the exercise.[31]

Four key lessons were learned from the distributed play and tabletop exercises in 2019. First, utilities and reliability coordinators must review their grid restoration and crisis management plans, particularly in support of national security. Second, critical communications facilities, such as fiber-optic backbones and wireless hubs, should be documented in grid restoration plans. Third, key supply chain elements (e.g., programmable logic controls, diffused gas analyzers, and gas turbines) should also be identified, and shared inventory management programs for the most critical component should be considered as well. Finally, as indicated in previous exercises, industry and government coordination between the United States and Canada still needed to be improved.[32]

GridEx VI, *2021*

The most recent *GridEx* better aligned scenarios across multiple utilities and interdependent critical infrastructure sectors. It was conducted in a largely virtual format; while prompted in part by the COVID-19 pandemic, this format more closely reflects how industry and government would respond in the real world. The scenario saw significant effects at each interconnection point between critical infrastructures, with effects to telecommunications and natural gas impeding the generation and transmission of electricity. Significant cyber and physical attacks further disrupted reliability and frustrated restoration efforts.[33]

The exercise continued the focus on operationalizing redundant communications pathways, clarifying the roles and responsibilities of operators with system-wide visibility, consultation with DOE on emergency orders, enhanced coordination with natural gas and telecommunications, and improved near-real-time consultation between the US and Canadian governments on emergency orders.[34] Though these conversations were similar to previous *GridEx* topics, the operational focus demonstrated process progress on several difficult issues, and closer interaction between interdependent sectors through the DHS Cybersecurity and Infrastructure Security Agency's Joint Cyber Defense Collaborative or DOE's Energy Threat Analysis Center.

Furthermore, participation in the November 2021 *GridEx VI* brought industry to a heightened level of preparedness and coordination. These valuable capabilities and relationships were in turn leveraged when Russia invaded Ukraine in February 2022. Once again, reality imitated the exercise (albeit in Eastern Europe). Preparation through *GridEx* and real-world events brought the North American electricity industry to a heightened state of vigilance, collaborating with the Joint Cyber Defense Collaborative to ensure a strong "shields up" posture for critical infrastructure.

THE NEED FOR COLLECTIVE DEFENSE

Unfortunately, real life continues to imitate the exercise, as demonstrated not only by events in Ukraine but also the Colonial Pipeline hack in May 2021. This ransomware attack disrupted the motor and jet fuel supply in the Southeastern United States—not the electrical grid, but too close for comfort. While electricity was not targeted, and no power outages occurred due to this event, the electricity industry quickly shared information in the aftermath, just as it had trained to do in *GridEx*. The natural gas pipeline system, which fuels electricity generation across North America, differs from the motor and jet fuel systems interrupted by the Colonial Pipeline hack. Nevertheless, the concept of reliance on a limited number of pipelines to provide critical generation supply caught the attention of industry, and this vulnerability became a key discussion topic for *GridEx VI* in 2021.

Similarly, after the SolarWinds hack in 2020, 25 percent of E-ISAC members reported having some version of the compromised software installed on their systems.[35] For the Microsoft Exchange compromise discovered in 2021, every utility using the compromised platforms had to scramble to patch its systems before malicious actors could exploit the vulnerability.[36] These events show that cyber threats continue to grow as adversaries look to gain access through vulnerable information technology supply chains.

As has been clearly demonstrated, *GridEx* has helped prepare the electricity industry for the evolving threat environment. Admittedly, some of the seemingly straightforward lessons learned—about the importance of communication and coordination, for example—are easier to identify through exercises and wargames than they are to resolve in real life. Simple solutions to complex problems are few and far between, especially when dealing with the over 3,300 stakeholders that make up the North American electricity industry. However, the consequences of not dealing with cyber and physical threats are stark. Kim

Zetter describes these threats as a "digital Pandora's box," opened by the Stuxnet attack: "The digital weapon didn't just launch a new age of warfare, it altered the landscape for all cyberattacks, opening the door to a new generation of assaults from state and nonstate actors that have the potential to cause physical damage and even loss of life in ways never before demonstrated."[37]

Nevertheless, by surfacing these issues (if need be, repeatedly year after year), exercises like *GridEx* can help organizations confront them, rather than sweep dangerous problems under the proverbial rug. For the utility companies that constitute critical infrastructure, as with the other kinds of businesses discussed in chapter 11, well-designed training and educational games can be a useful force for organizational change.

Participation in *GridEx* is voluntary. Fortunately, the value added is apparent to the players, as participation has grown from 75 utilities in 2011 to more than 520 in 2019 (there was a decrease to 293 in 2021 due to COVID-19 restrictions), with significant increases in small and medium-sized utilities, as well as state and local governments. *GridEx* occupies a central place in the critical infrastructure exercise calendar, and players report that they look forward to participating—potentially differentiating this event from sometimes-tedious command post exercises. *GridEx* provides a useful model for training and education in other sectors of critical infrastructure. It will continue to evolve to address evolving threats to the grid.

Responding effectively to cyber and physical threats against critical infrastructure requires collective action across the public and private sectors, in North America as elsewhere. The US government has taken notice of these threats, highlighting defense of critical infrastructure from cyber threats as the first pillar of the 2023 US National Cyber Strategy.[38] Including principal-level government officials in wargames like *GridEx* have helped socialize this idea among senior decision-makers. The need for collective defense is all the more acute today because many of the interdependencies and vulnerabilities that attackers can now exploit were not envisioned when the electric grid was first built. By helping practitioners and leaders "think the unthinkable," or at least the uncomfortable, about potential attacks and train for how they plan to respond, *GridEx* makes the industry and government more reliable, secure, and resilient.

NOTES

1. James B. Robb, "Keeping the Lights On: Addressing Cyber Threats to the Grid, Testimony to the US House Committee on Energy and Commerce Subcommittee on Energy," Washington, July 12, 2019, 2.
2. Avril Haines, "Annual Threat Assessment of the US Intelligence Community," Office of the Director of National Intelligence, April 9, 2021, 5.

3. David Nevius, "The History of the North American Reliability Corporation: Helping Owners, Operators, and Users of the Bulk Power System Assure Reliability and Security for More than 50 Years," North American Electric Reliability Corporation, 2019, 116-117.

4. William Jefferson Clinton, "Presidential Decision Directive/NSC-63," White House, May 22, 1998.

5. Robert McMillan, "Siemens: Stuxnet Worm Hit Industrial Systems," *Computerworld*, September 14, 2010, www.computerworld.com/article/2515570/siemens--stuxnet-worm-hit-industrial-systems.html.

6. Andy Greenberg, "Feds Uncover a 'Swiss Army Knife' for Hacking Industrial Control Systems," *Wired*, April 13, 2022, www.wired.com/story/pipedream-ics-malware.

7. North American Electric Reliability Corporation (NERC), "2011 NERC Grid Security Exercise After Action Report," March 2012, 7.

8. NERC, "Grid Security Exercise (GridEx) II: After Action Report," March 2014, 7.

9. Rebecca Smith, "Assault on California Power Station Raises Alarm on Potential for Terrorism," *Wall Street Journal*, February 5, 2014, www.wsj.com/articles/SB10001424052702304851104579359141941621778.

10. April Rubin, Livia Albeck-Ripka, and Matt Stevens, "North Carolina Power Outages Caused by Gunfire at Substations, Officials Say," *New York Times*, December 4, 2022, www.nytimes.com/2022/12/04/us/power-outages-north-carolina.html.

11. Cybersecurity & Infrastructure Security Agency, "ICS Alert (IR-ALERT-H-056-01): Cyber-Attack Against Ukrainian Critical Infrastructure," February 25, 2016 (revised July 20, 2021), www.cisa.gov/uscert/ics/alerts/IR-ALERT-H-16-056-01; Andy Greenberg, *Sandworm: A New Era of Cyberwar and the Hunt for the Kremlin's Most Dangerous Hackers* (New York: Penguin Random House, 2019), 109–15.

12. Greenberg, *Sandworm*, 141.

13. Michael Assante, Robert M. Lee, and Tim Conway, "ICS Defense Use Case No. 6: Modular ICS Malware," SANS Institute and E-ISAC, August 2, 2017, 1–3.

14. Cybersecurity and Infrastructure Security Agency, "Russian State-Sponsored and Criminal Cyber Threats to Critical Infrastructure," April 20, 2022 (revised, May 9, 2022).

15. Jenna McLaughlin, "A Digital Conflict between Russia and Ukraine Rages On Behind the Scenes of War," National Public Radio, June 3, 2022.

16. NERC, "Grid Security Exercise (GridEx) IV: Lessons Learned," March 2018, 13–17.

17. US Department of Energy, "Electric Disturbance Events (DOE-417)," no date, www.oe.netl.doe.gov/oe417.aspx; NERC, "CIP-008-6: Cyber Security—Incident Reporting and Response Planning," no date.

18. Electricity Subsector Coordinating Council, "Cyber Mutual Assistance (CMA)," no date, www.electricitysubsector.org/CMA/; *Critical Infrastructure Electric Security* §8240-1-1, (a)(7): "Grid Security Emergency."

19. NERC, "GridEx VI Lessons Learned Report," April 1, 2022, 8–12.

20. Cybersecurity and Infrastructure Security Agency, "ICS Joint Security Awareness Report (JSAR-12-241-01B): Shamoon/DistTrack Malware," October 16, 2012 (revised July 20, 2021).

21. NERC, "GridEx II," 17.

22. NERC, "Grid Security Exercise: GridEx III Report," March 2016, 10.

23. NERC, "GridEx III Report," 12–14.

24. NERC, 14–15.

25. NERC, "GridEx IV: Lessons Learned," v–viii.

26. NERC, 14.

27. NERC, vi.

28. NERC, 16–18.

29. Holden Mann, "NERC: COVID-19 Is Chance to Test GridEx Lessons," *RTO Insider*, March 31, 2020, www.rtoinsider.com/articles/19052-nerc-covid-19-is-chance-to-test-gridex-lessons.

30. NERC, "GridEx V Lessons Learned Report," March 2020, 2–6.

31. NERC, 1.

32. NERC, 7–11.

33. NERC, "GridEx VI: Lessons Learned Report," 5–7.

34. NERC, 8–12.

35. Robert Walton, "NERC Finding 25% of Utilities Exposed to SolarWinds Hack Indicates Growing ICS Vulnerabilities, Analysts Say," *Utility Dive*, April 15, 2021, www.utilitydive.com/news/nerc-finding-25-of-utilities-exposed-to-solarwinds-hack-indicates-growing/598449/.

36. Microsoft Threat Intelligence Center, "HAFNIUM Targeting Exchange Servers with 0-Day Exploits," March 2, 2021, www.microsoft.com/security/blog/2021/03/02/hafnium-targeting-exchange-servers/.

37. Kim Zetter, *Countdown to Zero Day: Stuxnet and the Launce of the World's First Digital Weapon* (New York: Penguin Random House, 2014), 374.

38. Joseph R. Biden, "National Cybersecurity Strategy," White House, March 1, 2023, 7.

BREACHING THE C-SUITE: WARGAMING IN THE PRIVATE SECTOR

MAXIM KOVALSKY, BENJAMIN H. SCHECHTER, AND LUIS R. CARVAJAL-KIM

Malaise Pharmaceuticals is having a bad day.[1] Its CEO, Morgan DeSanto, worked for years to establish credibility with investors and grow the stock price. But the last three years of gains have just been wiped out in a matter of minutes. Morgan is pacing up and down the boardroom–turned–crisis response center. "Why isn't HR here yet?" she asks Tim Matthews, the company's chief information security officer (CISO). Although Tim had been on the job for over a year, this is his first time meeting Morgan. Not for lack of effort: he repeatedly tried to present his cyber risk assessment findings for several months. "Stacey is on the first morning flight," interjects Mike, Morgan's chief operating officer (COO). It has been four hours since hackers posted correspondence about sensitive research stolen from the company on social media; someone needs to take charge. "I suppose we'll have to add 'crisis commander' to the job description for my replacement," thought Mike.

The chief legal and investor relations officers speak in raised voices at the other end of the room, managing a tsunami of incoming calls from the press, investors, and—ominously—federal regulators. Mike is following a stream of questions on his phone. Just as he is about to issue critical guidance on the fast-moving situation, his email app displayed a connection error. "Attention!" yells Tim. "My team thinks email is compromised. I had to make the call to contain the damage. We should be getting it back online shortly."

The previous day, someone with the handle *"killbigpharma"* tweeted at Malaise Pharmaceuticals: "Come clean now or be exposed as a greedy fraud." The message was deemed to be a trolling attempt and was ignored. However, if someone had investigated the handle, they would have seen the user leaking another pharmaceutical company's internal memos several months earlier in the middle of a price-hiking scandal. Twenty-four hours later, *killbigpharma* posted internal emails from one of Malaise's chief researchers about a breakthrough drug due for impending approval from the Food and Drug Administration. In those emails, the researcher raised concerns about being pressured to misrepresent recent findings from clinical trials.

The neatly stacked pieces of daily corporate life at Malaise have been upended as a result. The psychological pressures of fear and uncertainty are overwhelming, leading to a near-complete collapse in communication within the executive team. No one knows exactly who is in charge or what is happening—whether these tweets were the beginning or the end. Yet everyone seems to have different opinions about what to do next. Tim, the company's CISO, attempts to get the room's attention with an update: "The domain controller was compromised a week ago. It is safe to assume *killbigpharma* is sitting on a huge pile of our data." This is indeed only the beginning.

WARGAMING FOR BUSINESS RESILIENCE

The crisis consuming the fictional Malaise Pharmaceuticals is not unique. Organizations of every type and size should expect to grapple with cyber incidents that escalate into business crises.[2] When viewing cyberspace through the lens of people, processes, and technology, many large enterprises are prepared to deal with the technical aspects of a cyber incident. However, large or small, it is not uncommon for executive teams to be less organized when the event escalates beyond "ones and zeros" and affects critical business functions.[3] They are often ill equipped for the people and process aspects.[4] Is there a plan? And are people prepared to execute that plan? Despite numerous high-profile cyberattacks over the years, the executive suite's disinterest in cybersecurity issues is well documented.[5]

The Malaise Pharmaceuticals vignette highlights a fundamental problem: the limited appreciation of how cybersecurity affects an organization's resilience, namely, "the ability to prepare for and adapt to changing conditions and withstand and recover rapidly from disruption. Resilience includes the ability to withstand and recover from deliberate attacks, accidents, or naturally occurring threats or incidents."[6] This is seen in at least two problems highlighted in the vignette. First is a lack of coordination between cyber and noncyber functions, both during "business as usual" and then during a crisis. Second, the executive team in this scenario is not responding to the crisis in a unified or coordinated manner. The common thread across these problems is an executive team unprepared for a rapidly escalating cyber incident.

Wargames are useful outside national security contexts, helping leaders and organizations better prepare for today's uncertain and insecure world.[7] Breaches will happen. Organizations should face them with clarity of how they will respond. Private-sector cyber wargames are well suited for solving cybersecurity problems, especially when people—not necessarily technology—require support.[8]

A well-constructed cyber wargame helps organizations avoid the type of chaos consuming Malaise Pharmaceuticals by adhering to two important truths:

- Truth 1: Effective cyber wargames go beyond presenting cyber incidents as just technical problems by translating the technical actions and effects of a cyber event into business-relevant terms.
- Truth 2: Effective cyber wargames help increase an organization's resilience to cyber threats and incidents by affecting organizational change.

This chapter explores cyber wargames as instruments to drive change and improve cyber resilience. The first section explores the history and typologies of private-sector wargames. The second section discusses the most common problems for private organizations, providing basic context. The third and fourth sections examine how to diagnose and frame these problems and address forces that might otherwise hinder the effectiveness of cyber wargaming for the c-suite. The chapter concludes with reflections on driving change and the implications for cyber wargames and technology wargaming more broadly.

WARGAMES AND CYBER WARGAMES IN THE PRIVATE SECTOR

The private sector has diverse needs, ranging from those of startups and small family businesses to giant multinational corporations. The wargames used by private-sector firms tend to reflect this reality, varying widely in their objectives and designs.[9]

Private sector wargames, sometimes called tabletop exercises or crisis simulations, date back to the early 1950s.[10] Cyber risk management has a history reaching back to the late 1970s, although terms such as "information security," "information assurance," and "cybersecurity" would come later.[11] Despite these long legacies, cyber wargames in the private sector are relatively new—especially wargames that grapple with cybersecurity instead of routine technology adoption and modernization challenges.[12] Intel pioneered the first known private-sector cyber wargames in 2008, inspired by *Digital Pearl Harbor*, a cyber wargame held in 2002 at the US Naval War College, which is discussed in chapter 3.[13]

One way to classify business wargames is to broadly divide them into two categories: games for preparing and games for responding.[14] The first category of games helps organizations with strategy and planning, providing insights to improve decision-making and resource allocation at the strategic level. They usually focus on testing and refining plans and strategies, understanding competition dynamics, the effects of major business decisions, and changing markets. These wargames fit relatively easily into an organization's planning cycle.

The second category of games, and the focus of this chapter, are for crisis response—wargames that help companies not only plan for but also manage a crisis. Granted, there are other typologies for business wargames. For example, another distinction is between general business games, which focus on competition and decision-making, and functional games, which focus on specific business functions.[15] That said, we argue that crisis response games warrant special attention, since they help companies test and improve existing processes and prepare employees for the stress of an actual crisis. When well executed, these games help identify important problems, which can serve as a wake-up call for complacent leaders.

Wargames intended to drive change and improve cyber resilience are predominantly response-oriented. These games are designed to improve incident response by helping distill a clear idea of the organization's problems and what should be done. In many circumstances, using wargames to *sell an idea* can—and should—be considered a problem, or at least viewed with considerable skepticism.[16] However, if the goal of the wargame is to generate change, selling or socializing new ideas is not only acceptable; it is also explicitly its purpose. In these instances, players experience the game, internalize its lessons, and implement them within the organization. When those players are c-suite executives, they are particularly well positioned to drive change in the real world, based on the lessons learned in the game. To run these cyber wargames successfully, the designers must have extensive knowledge of the problems, organizations, and possible solutions; they must also engage the participation of leadership and other relevant agents of institutional change.

THREATS TO CYBER RESILIENCE

Designing a cyber wargame to improve organizational cyber resilience requires diagnosing the problem. Putting aside the purely technical aspects and focusing on the people and processes, problems often fall into three broad categories: governance (how decisions are made and who gets to make them); processes (the tasks and activities for accomplishing objectives); and communications (sharing information within and outside an organization). These categories are interlinked, with effective processes and communication both being essential parts of governance. Problems rarely fit neatly into one category, and are a complex mix of intersecting issues that may undermine an organization's cyber resilience.

Governance

Governance is the structure to control and direct an organization. Who is empowered to make decisions? When do they make them? And how? While seemingly obvious, these are not clear-cut questions during a crisis, and even less so during a cyber incident. Is the responsible authority the COO, chief information officer (CIO), or perhaps the CISO?[17] For example, the decision to take information systems offline may have profound implications for a particular business line or the enterprise as a whole. To further complicate these questions, some subordinates and stakeholders outside the c-suite should be consulted before such a decision, as well as informed after a decision is made.

Practical governance questions are also complicated because leadership is a deeply human affair, with all its messy implications. Wargames can bring to light unexpected decisions companies may need to face, such as whether to shut down parts of the corporate network or client-facing systems. In times of crisis, fear and uncertainty about the organization's future get tangled with concerns about personal job security and future career prospects (e.g., the "games within games" described in chapter 7 can mirror those in real life). Under duress, even leaders who have demonstrated strategic vision and operational prowess may become unable to take decisive action. As a result, decision-making may become disorganized or completely stalled. At the opposite end of the spectrum, functional leaders with a bias for action may make calls without properly considering their second- and third-order effects, which may be well known by leaders in other business functions.

A good incident response plan can bolster governance during a crisis and support effective business processes and communications. Every employee, from the c-suite to the cybersecurity team, should understand the range of decisions they are empowered to make in times of crisis—including those decisions where consultation with other team members is necessary and those that should be escalated to higher leadership. Such a plan can help individuals better understand their roles by reaffirming (or reasserting) governance structures during a crisis. However, even a good incident response plan is not a panacea.

Processes and Communication

Process and communication are critical in their own right, and are nuanced components of governance. Processes include activities that allow an organization to implement executive decisions and accomplish its objectives. Communication is an organization's ability to transmit information clearly and effectively, including guidance from staff members ranging from executives to subordinates, between business units, within business units, and from external stakeholders and customers. Even if the governance structure is sound, communication and the processes responsible for implementation may be flawed, undermining an organization's ability to manage and quickly recover from a cyber crisis. In the example of Malaise Pharmaceuticals, the urgent decision to take an information system offline should be communicated to the right teams simultaneously, and those teams must have a process to implement that decision safely.

A cyber incident response plan is "a predetermined set of instructions or procedures to detect, respond to, and limit consequences of malicious cyber attacks against an organization."[18] Ideally, such a plan identifies the necessary processes for managing a crisis. However, while necessary, a plan is insufficient; it must also have been exercised and updated accordingly. Wargames help determine what exists in practice versus what exists on paper. They can help test an existing, well-rehearsed cyber incident response plan; motivate improvement if an existing response plan is poorly developed or integrated; and serve as a call to action if no plan exists.

Even if an incident response plan exists, implementing it will require effective communication. During a crisis, an unprepared executive team may not fully appreciate the need for rapid, disciplined communication. Risks emerge when the c-suite cannot coordinate succinctly to provide clear communication with subordinates, leading to inaction or, worse yet, letting confusion reign. Additionally, a failure to understand and engage relevant stakeholders, from the regulators, to shareholders and the general public, risks amplifying the damage. Effective crisis response requires coordination with business operations, public relations, legal counsel, and, in dire situations, physical security.

DIAGNOSING THE PROBLEM

Designing wargames that motivate change and improve cyber resilience requires substantial and specialized preparatory work, some of which may seem out of place in traditional wargames. Successful wargame design relies on knowing the organization's challenges, both internal and external. Sometimes, outside threat actors are the greatest risk. Other times, insider threats or even the c-suite itself may be the greatest liability. Regardless, running a successful game requires successfully diagnosing the problem.

Know Your Enemy

As Sun Tzu argues in *The Art of War*, "If you know neither the enemy nor yourself, you will succumb in every battle. If you know yourself but not the enemy, for every victory gained, you will also suffer a defeat. If you know the enemy and know yourself, you need not fear the result of a hundred battles."[19]

The success of a c-suite cyber wargame hinges on deep engagement by the c-suite participants. The game must resonate with the players. Maintaining a sense of realism throughout the exercise is fundamental; the relevance and credibility of events within the game help participants invest in the wargame and its outcome.

Good intelligence about potential adversaries provides realism and the basis for the credibility of the exercise. Cyber threat intelligence is empirical knowledge about the threat actors' intent and capability to harm the organization. This intelligence is aggregated from diverse sources, including past incidents at the organization or its peers, government bulletins, and other sources. The goal is to create a compelling notional adversary in the wargame that represents the threat actors the organization may face. This includes realistic behavior; motivations; objectives; capabilities in terms of techniques, tools, and procedures; and communications with the target organization and general public.

Know Your Organization

While knowing the enemy is one side of the coin, knowing yourself is the other. Credible threats should be presented with credible vulnerabilities—intelligence about external actors should be overlaid against vulnerabilities likely present within the organization. Trusted agents are needed here to act as brokers between the wargame designers and the organization's cyber defense team to help identify gaps in visibility, a lack of patching on systems intolerant of downtime, and weak compensating controls. While crisis response wargames need not be technical, they must be technically sound.

Designers must understand the organization's processes, systems, culture, and pathologies. For example, what are the organization's most important systems that would pose an existential threat to the firm if degraded or rendered inoperable? This is done concurrently with gathering cyber threat intelligence; the internal aspects of an organization will influence the types of threats it faces. Finally, designers must understand the people—their working styles and equities.

The wargame tests the extent to which, during a stressful event, a group of individuals can come together as a team and transparently communicate and collaborate to resolve the crisis by tapping into each other's strengths. Positive outcomes during a crisis often hinge on the strength of relationships and trust among the executive team. Therefore, it is important to examine the working styles and habits of the participants to determine what may need to change to arrive at the optimal collective working style. On one hand, does a particular individual on the leadership team have a strong bias for action? On the other hand, does someone else prefer to consider everyone's perspective before making or recommending a decision? Are the participants averse to second-guessing or do they prefer to course-correct as new information becomes available?

The organization will gain the most value from a cyber wargame or executive tabletop when the leadership team's working styles are incorporated into the exercise design. The desired outcomes of this process may include exposing working styles and habits that are counterproductive during a crisis, or ensuring that leaders understand the importance of adjusting to their peers', subordinates', or superiors' working styles.

Accounting for these people is an important difference between wargaming for the c-suite and many other professional games; the game is for these *specific* players. Therefore, effective games must properly characterize the cyber issues they face and resonate with the key players by ensuring their equities are affected and exercised during the simulated crisis.

FAIL CONDITIONS FOR C-SUITE GAMES

Wargames go astray for many reasons.[20] Common reasons include lack of an appropriately senior sponsor or "champion" of the change, failure to acknowledge the need for the change, and excessive delegation to staff without follow-up or accountability for implementation. Before going down this road, organizations should consider a few particularly acute issues for cyber wargames at the c-suite level, specifically buy-in, attention, and the wrong design team.

Buy-In and Cooperation

Several challenges stand out concerning buy-in and cooperation from wargame participants. Tacit disagreement with the wargame's stated objectives may lead to attempts by key stakeholders to influence the design of the exercise. If left unacknowledged, these players may end up "fighting the scenario" during the event or display a lack of interest.

Players may also disagree on the need to invest time and resources in the exercise against other perceived priorities. While they may not be actively resisting the premise of the event, their attention and participation may be lacking, leading to an inability to integrate valuable perspectives into game design.

Finally, the game's objectives may be misunderstood. Even without ill intent, misunderstanding can result in misalignment between inputs and outcomes, which can lead to frustration on the part of players during the design process and the event itself.

Attention and Ethics

In an ideal world, the timelines align and the c-suite is interested. But that still does not ensure they can fully engage. The same reasons that cybersecurity often gets shortchanged—lack of understanding and prioritization—can undermine cyber wargames. The problems of the "here and now" often take precedence over the problems of tomorrow.

Getting the attention and trust of the c-suite can be difficult if the wargame's ultimate purpose is hidden from some participants. This can be a challenge when fully understanding the game's objectives can influence participants' behavior to such an extent that these objectives cannot be achieved. Such "hidden" objectives may include an assessment of a new leadership team by the board, for example, highlighting the need to reorder power structures, or other tests and experiments that may produce artificial results if widely known. Some of the ethical issues that arise when wargames are used for research also apply when they are used to educate, socialize, or sell ideas. Attention should not be purchased at the cost of harmful deception.

The Wrong Design Team

Many wargames are not designed by an internal team dedicated to such events. More often than not, businesses and other organizations outsource this task to an external consultant. Yet even when objectives are clearly communicated, some design teams may lack sufficient expertise and experience to achieve the desired outcomes.

For example, suppose the design team is accustomed to developing exercises that test the technical competencies and tactics of cyber defenders. In this case, the resulting wargame may be too technically "deep in the weeds" at the expense of the higher-level objectives that are of greatest concern to the c-suite. Particularly when working with external consultants, sponsors and designers must work together to avoid building wargames that are inadvertently "designed to fail."

DRIVING CHANGE

Cyber wargames can be vehicles for change. Like all vehicles, they require a driver to be effective. While practice and training have value, we argue that many cyber wargames are ultimately intended to affect change. This is realized through designing exercises to inform, educate, and *persuade* participants and observers of a particular position or concept to help springboard organizational change. Critics may argue that using wargames to subtly or subliminally influence an audience to act is inappropriate. However, if organizations do not evolve or adapt as a result of cyber wargaming, then the exercise is probably a missed opportunity.

It is tempting to treat organizations as monolithic entities when trying to drive change. But this overlooks all the individuals involved. Effective wargame design often focuses on initiating change at the individual level, consistent with theoretical frameworks for education, such as Malcolm Knowles's work on adult learning.[21] Using this model, we know that effective cyber wargame design should support individuals synthesizing and learning on their own, with a facilitator as a guide and not a lecturer; should focus on participants learning through experiences, including mistakes; should assume that individuals both want to change and derive personal or professional benefit; and should recognize that individuals can change and have the requisite training and skills needed to change.[22]

Helping people become change agents can provide the critical mass needed to change organizational culture. The basis for these changes, whether for an individual or an organization, rests on people, processes, and technology.

People

People include what they know, believe, and are prepared or trained to do. Many organizations enter a cyber wargame believing that the people on their executive team can manage the most straightforward cyber incidents. Well-designed, realistic scenarios demonstrate that many cyber incidents are anything but straightforward. These games highlight the human dimension. Organizations can become overwhelmed because leaders neglect the challenges of individuals and teams who may be resource constrained (limited staff and budget), skill constrained (niche or highly complex challenges), and work or life-constrained (team members have to rest and recharge). The wargame drives these realities to the surface, helping teams to take stock.

Process

Processes include what people do, how they do it, and what policies or plans organizations of people have in place. Organizations often go through great efforts to establish and maintain technical controls. Cyber wargames can help illuminate instances where a skilled attacker can circumvent technical controls—often by exploiting human factors. One common lesson comes from the risks of automating broken processes, which result in "garbage in, garbage out" feedback loops for organizational processes. Whatever the reason for these risks, being aware of them and their consequences can catalyze change.

Technology

Technology includes the tools people use and need. Information technologies are too often treated as a panacea for managing cyber risks with the mistaken belief that "the tool will fix it." Advocating for change around tools can also be troublesome because of the high cost associated with buying and deploying new technology, and managing associated organizational changes. People can likewise become heavily invested in the "success" of their chosen tools in the environment, even if it is a pyrrhic victory. Cyber wargames can help open the conceptual aperture around new and legacy security tools. Technology can help enhance an organization's resilience, but it cannot in and of itself unilaterally address risks without appropriate people and processes in the loop.

CONCLUSION

Cybersecurity has been a critical part of business resilience for decades. Malaise Pharmaceuticals is illustrative of the high stakes of cyber incident response. The human decisions made by senior leaders can make or break an organization when it is rocked by a cyber crisis. A well-executed cyber wargame can prepare the c-suite for that crisis, helping it respond, contain, and recover more successfully.

Cyber wargames target the human dimension of cybersecurity. Their purpose is to improve an organization's cyber resilience by demonstrating the consequences of weak governance, poor process, and faulty communications. Designing cyber wargames requires a deep understanding of the organization and the people who constitute it. Effective design serves to create a bespoke wargame tailored to the organization's specific needs.

There are opportunities to improve cyber wargaming for executives through additional research. For example, private-sector wargames would benefit from further integrating interdisciplinary research from education, sociology, and psychology. New methods and tools to identify personality types, leadership styles, group dynamics, and organizational characteristics and to link them to wargame design would lead to better learning. Prioritizing long-term measures and assessments of wargame outcomes—specifically institutional and behavioral change over multiple games—would also help refine and improve existing methods for the c-suite.

Finally, cyber wargames can help drive change to increase cyber resilience. They could be used in any organization with an executive team, governance structures, and an interplay of people, processes, and technology. Wherever they are used, the ethics of player deception and persuasion warrant serious consideration. The same is true when wargames are used to help leaders consider the potential consequences of other emerging technologies for their organizations. The only overarching requirements are a sizable human dimension, a need to improve resilience, and a call for decisive change.

NOTES

1. The views expressed in this chapter are those of the authors alone and do not necessarily represent those of the US Navy, US Department of Defense, US government, or any other institution or business.
2. Yoram Golandsky, "Cyber Crisis Management, Survival or Extinction?" paper presented at 2016 International Conference on Cyber Situational Awareness, Data Analytics and Assessment (CyberSA) 2016, 1–4; Council of Economic Advisers, "The Cost of Malicious Cyber Activity to the US Economy," 2018, www.whitehouse .gov/wp-content/uploads/2018/03/The-Cost-of-Malicious-Cyber-Activity-to-the -U.S.-Economy.pdf.
3. Charles Aunger, "Cybersecurity Is Not Necessarily a Technology Issue," *Forbes*, October 30, 2019, www.forbes.com/sites/forbestechcouncil/2019/10/30/cybersecurity -is-not-necessarily-a-technology-issue/.
4. C. Inglis, "Cyberspace: Making Some Sense of It All," *Journal of Information Warfare* 15, no. 2 (2016): 17–26.
5. Melissa E. Hathaway, "Leadership and Responsibility for Cybersecurity," *Georgetown Journal of International Affairs*, 2012, 71–80; Louis Columbus, "Cybersecurity's Greatest Insider Threat Is in the C-Suite," *Forbes*, May 29, 2020, www.forbes .com/sites/louiscolumbus/2020/05/29/cybersecuritys-greatest-insider-threat-is-in -the-C-suite/.
6. Joint Task Force, "Risk Management Framework for Information Systems and Organizations (Revision 2)," NIST Special Publication 800-37, 2018.
7. Benjamin Gilad, *Business War Games: How Large, Small, and New Companies Can Vastly Improve Their Strategies and Outmaneuver the Competition* (Franklin Lakes, NJ: Red Wheel Weiser, 2008); Peter P. Perla and John Curry, *Peter Perla's the Art of Wargaming: A Guide for Professionals and Hobbyist* (London: Lulu, 2011); Tucker Bailey, James Kaplan, and Allen Weinberg, "Playing War Games to Prepare for a Cyberattack," *McKinsey Quarterly*, 2012, 1–6.
8. Benjamin Schechter, "Wargaming Cyber Security," *War on the Rocks*, September 4, 2020, https://warontherocks.com/2020/09/wargaming-cyber-security/.
9. David Ormrod and Keith Scott, "Strategic Foresight and Resilience Through Cyber-Wargaming," in *Proceedings of ECCWS 2019 18th European Conference on Cyber Warfare and Security* (London: Academic Conferences and Publishing, 2019), 319.
10. Bernard Keys and Joseph Wolfe, "The Role of Management Games and Simulations in Education and Research," *Journal of Management* 16, no. 2 (June 1990): 307–36; David C. Lane, "On a Resurgence of Management Simulations and Games," *Journal of the Operational Research Society* 46, no. 5 (1995): 604–25; Anthony J. Faria et al., "Developments in Business Gaming: A Review of the Past 40 Years," *Simulation & Gaming* 40, no. 4 (2009): 464–87.
11. Michael McShane, Martin Eling, and Trung Nguyen, "Cyber Risk Management: History and Future Research Directions," *Risk Management and Insurance Review* 24, no. 1 (March 2021): 93–125.

12. For one review, see Giddeon N. Angafor, Iryna Yevseyeva, and Ying He, "Game-Based Learning: A Review of Tabletop Exercise for Cybersecurity Incident Response Training," *Security and Privacy* 3, no. 6 (2020).

13. Thomas C. Greene, "Mock Cyberwar Fails to End Mock Civilization," *Register*, 2002, www.theregister.com/2002/08/30/mock_cyberwar_fails_to_end; Tim Casey and Brian Willis, "Wargames: Serious Play That Tests Enterprise Security Assumptions," Intel Corporation, 2008; Intel Corporation, "Playing Wargames in the Enterprise," Intel Newsroom, 2010, https://newsroom.intel.com/editorials/playing-wargames-in-the-enterprise/.

14. Daniel F. Oriesek and Jan Oliver Schwarz, *Business Wargaming: Securing Corporate Value* (Burlington, VT: Gower, 2008), 41–42, 73.

15. Gerhard R. Andlinger, "Business Games-Play One," *Harvard Business Review* 36, no. 2 (1958): 115; Kalman J. Cohen and Eric Rhenman, "The Role of Management Games in Education and Research," *Management Science* 7, no. 2 (1961): 131–66.

16. Stephen Downes-Martin, "Exploit Small Group Dynamics During Analysis to Support the Decision You Want," paper presented at 13th NATO Operations Research and Analysis Conference, Ottawa, 2019; Christopher A. Weuve et al., "Wargame Pathologies," Center for Naval Analyses, 2004; Schechter, "Wargaming Cyber Security."

17. Deloitte Consulting LLP, "Tomorrow's COO: A Changing Role in a Changing World," no date, www2.deloitte.com/content/dam/Deloitte/us/Documents/financial-services/us-fsi-tomorrows-coo-a-changing-role-in-a-changing-world.pdf.

18. Marianne M. Swanson et al., "Contingency Planning Guide for Federal Information Systems," November 11, 2010.

19. Sun Tzu, *The Art of War* (New York: Oxford University Press, 1971).

20. Weuve et al., "Wargame Pathologies."

21. Malcolm S. Knowles, *The Modern Practice of Adult Education: From Pedagogy to Andragogy*, revised and updated (Cambridge: Cambridge Adult Education, 1980); Malcolm S. Knowles, Elwood F. Holton, and Richard A. Swanson, *The Adult Learner: The Definitive Classic in Adult Education and Human Resource Development*, 7th ed (Boston: Elsevier, 2011).

22. "Adult Learning Theories," US Department of Education, 2011, https://lincs.ed.gov/sites/default/files/11_%20TEAL_Adult_Learning_Theory.pdf.

12

DEVELOP THE DREAM: PROTOTYPING A VIRTUAL CYBER WARGAME

CHRIS C. DEMCHAK AND DAVID H. JOHNSON

Across established democracies, there is a shortage of folks able to understand and work with emerging information technologies. Many of these technologies—ranging from artificial intelligence (AI) to quantum computing—are what we call the "offspring" or the "children of cyber." Human talent and expertise for working with them are lacking.[1] Many government and industry leaders struggle even to understand the basic structure of cyberspace that serves as the substrate of our digital world.[2] Limited comprehension persists, despite countless compliance training sessions, lectures, exhortations, bad news stories, and putatively helpful policy briefs about cybersecurity.

One problem is the lack of easily accessible and interactive places—sandboxes, if you will—to learn about how cybered conflict could play out in real life.[3] Game-based learning can help educate students in military and civilian life alike. However, there are few virtual cyber wargames—namely, serious wargames played on a computer—that adequately capture the complexity, dynamism, and trade-offs of the emergent and dangerous realities of this domain. Filling this gap is essential. Such a game would include high fidelity, replay, automated adjudication, realistic representations of complex conflict, and immersiveness to encourage learning, testing, and creative responses.

Starting in 2016, we led a small team of researchers at the US Naval War College and the New England Institute of Technology in an attempt to fill this gap.[4] This chapter presents that work as an experiment in prototyping a virtual cyber wargame. We describe the development process in a style akin to a travel log. We also discuss the results of this experiment, noting lessons learned and future efforts that might build on this effort.

AMBITIOUS PLANS AND ALTERNATIVE PATHWAYS

Virtual cyber wargames were rare in 2016. The closest analogs were generally found in the form of highly technical "ranges" for training systems administrators, policy tabletop exercises that relied on face-to-face meetings and human adjudicators, and cobbled-together "cyber effects" that were bolted on after the fact to otherwise traditional wargames.[5] People could not "play through" an immersive cyber wargame to learn how to use cyber tools, operations, and campaigns. No one could readily replay a game, replicate a particular game sequence, test the reliability of adjudication results, or test new ideas later.

To address these issues, we sought to design a cyber wargame played on a networked computer that could provide sufficiently *high fidelity, repeatability,* and *standardized conditions* to constitute "practice," and thereby practical as opposed to theoretical education. We wanted a game that human players could both learn to handle and experiment with vis-à-vis the complexity of cyber conflict—especially "in the gray zone" to "the left of boom" on the spectrum of conflict, namely, before or below the threshold of conventional armed conflict.[6] That is where the bulk of cybered conflict occurs today. Adversaries use cyberattacks while seeking to stay below what the defenders will see as triggers requiring a kinetic response.[7] Furthermore, we wanted a serious game to mimic the experience of conflict where concrete effects are imposed by adversaries that can organize large campaigns, at any proximity or distance, using highly deceptive tools while obscuring their identity.[8] Our goal was a tool that could be used by dispersed and diverse audiences—students—concerned with understanding cybered conflict.

We considered four options: a tabletop exercise, a card game, a computerized simulation, and a virtual cyber wargame. Tabletop exercises are relatively easy to create and execute. They can also be made quite immersive with the right storyline and information to effectively create a sense of urgency.[9] Yet this modality lacks replay and automated adjudication that can weigh the same actions the same way in each instance of play. Furthermore, one can only play these games by gathering players together at more or less the same time.

Card and board games are often insufficient as well. Turn-based versions lack the simultaneity of actions and effects found in the real world.[10] In cyberspace, nobody waits their turn. Replay for review is likewise difficult with these games.[11] The lack of visual cues is another challenge, along with the complexity of rules. According to Jesse Schell, an early leader in commercial virtual game design, "a game like *Warcraft* could conceivably be a board game, but there would be so many rules to remember and state to keep track of that it would quickly become a dreary experience. By offloading the dull work of rules enforcement onto the computer, games can reach depths of complexity, subtlety, and richness that are not possible any other way."[12]

Simulations came closer to meeting our desired requirements. This methodology has a long history.[13] In principle, simulations can be coded to have a great deal of fidelity, replay, and automated adjudication.[14] However, as described in chapter 2, simulations lack human players and, similarly, immersiveness. The mechanics of automated simulations are not meant to draw in the variability, curiosity, attention, or passion of people. These are the aspects of immersion that induce us to return again and again to try something new and practice because the participant wants to.[15]

Only a virtual, "serious" cyber wargame offered us a solution for gamified computer simulations that involve human players.[16] In general, games amplify the players' enjoyment of the process, stimulating the challenge to return, replay, and try new strategies. A serious game adds the requirement for learning as education, or, as described earlier in this book, research and analysis. Virtual versions offer new opportunities for achieving a good deal of Bloom's seminal taxonomy of learning—namely, remember, understand, apply, analyze, evaluate, and create.[17] If the virtual game is designed well, the players will also be able to perform more advanced learning techniques: evaluating outcomes against preferences repeatedly, and creating new processes, mental models, points of view, and strategies. In addition, such games offer the possibility of simultaneous actions by players, thereby creating the unexpected surprises and innovations that can emerge in complex adaptive systems.

While a virtual game was best suited to our criteria, we could not find such a game in 2016. So, we decided to build one.

AMBITIOUS PLANS AND SCARCE RESOURCES

How can one make a game as close as possible to the ideal web-based, unclassified, portable, tailorable, single or multi player, replayable, internally adjudicated, high-fidelity, cybered conflict game? Our goal was a multiobjective game such that students taking a class or wanting to test hypotheses about decision-making would enjoy repeatedly playing. The bar was set high, and we expected the first few prototypes to fall short. But our project was borne of a deep desire to see if a small group of people could make the game happen.

Another practical consideration that drove much of the early prototyping was our desire for universality in potential uses. This was to be a general-purpose cyber wargame that could be tailored to fit multiple uses. It was also to be accessible by many users, remote or cloud-based, and asynchronous or face-to-face.[18] These multiple objectives and audiences added to the design and construction challenges. It is easier to have narrow objectives for only a small number of players. But then many advantages of diverse interactions, objectives, and players are lost.[19] Pushing the envelope in a multiobjective, multiagent game involved risk, and we accepted these risks given the potential rewards.

We were constrained by hard limits on funding and time. The project had an initial budget of about $25,000 to make a basic prototype. While the limited budget available at the Naval War College prevented us from hiring an experienced gaming company, equally important was our underlying desire to create a game owned by the Navy, for the Navy, and for use by the Department of Defense more broadly. Doing so would not only reduce the cost for teachers, students, and researchers to acquire and use the game (i.e., without needing to pay license fees to a third party) but also decrease the government's dependency on an external company for operating, sharing, and updating the game.

Obtaining coding talent with our limited budget was a critical challenge. Fortunately, the Naval War College is across Narragansett Bay from the New England Institute of Technology, which offers three-year degrees in game design and coding. Working in collaboration across the two campuses, our team started phase one of this project by identifying opportunities for student volunteers.

In August 2016, our effort began with a weekend "game jam" to identify the New England Institute of Technology's gaming students who were interested and able to contribute to building a virtual cyber wargame. Students formed teams with whomever showed up on the Friday afternoon and worked through Sunday. Those that persisted then presented their game design. Pizza and soft drinks provided fuel for this marathon session. Seven teams succeeded in roughly outlining creative designs during this game jam.

Phase two involved seeking a design for a complete prototype. Armed once again with food, the project team offered students a lecture on cybersecurity and an impassioned call to develop prototypes of their games by the end of the year. A winning design would then be chosen and purchased for $2,000. Forty-odd students signed up, breaking into roughly eight teams. We mentored these teams to completion, and eight designs were submitted in December 2016. No one single design satisfied our full vision, but three of them each offered something worth incorporating into the next phase.

Phase three was a five-month effort to produce the first working, albeit exceptionally basic, prototype combining the three designs. Four students from across the three winning design teams agreed to form a single team and take a small contract to create a working, single-player prototype. This was the first of several protypes, all of which were built using the 2017 version of the open source Unity engine. This was one of the most popular of the virtual game environments for designers to use to create backgrounds, scenes, and coding actions in games. This version of Unity was also the only one authorized for use on unclassified networks at the Naval War College.[20]

In May 2017, the students—graduating seniors, working between classes—delivered a prototype. Our expectations were modest. The game's mechanics worked. But the fidelity, graphics, complexity, and substantive depth that we ultimately desired were not yet there.

Phase four sought to refine a more robust prototype that focused on the Navy. Fortunately, we were able to secure additional summer support for two students, who spent several months expanding on the prototype. These students were both designers, with basic but not advanced coding skills. Thus, the work went slowly. The prototype improved, but many shortcomings still lingered.

Phase five aimed to refine the prototype into a more robust version able to support multiple players across the same server. With funding again challenging, the Naval War College's Center for Naval Warfare Studies agreed to support a single student to continue work as a contractor on the game. Delivery of a working game was estimated to be about twenty-four months. That timeline and funding proved adequate for an advanced prototype but not for a completed game, given the requirements and constraints described below.

The development process became increasingly knowledge intensive as the game became more Navy focused. The initial prototypes had been neutral with regard to what kind of military actors. Adding specific ship types and naval systems significantly increased the knowledge burden of gathering and then curating the ship and then fleet data to embed in the game. We redesigned the user interface to include key attributes of ships from different countries. Fleets of ships were renamed "task forces," with a command ship and a variety of specialized surface action groups. Their attributes included speed, costs, duration, and areas of operations, as well as educated guesses from open sources about the number of internal computerized systems for each ship. These data were embedded in the game code.

In addition to open source research, interviews with experts at the Naval War College helped us construct six general categories of mission critical ship systems (propulsion, weapons, logistics, support, navigation, communications) that might be harmed by cyber campaigns. Surface action groups and command ships each vary in resilience, battle capabilities, network complexity, communications systems, logistics capabilities, speed, humanitarian capabilities, endurance, and readiness.

Players' choices when creating their task forces created a distinct cyber "attack surface" for each task force combination. We coded six separate missions (freedom of navigation, humanitarian assistance / disaster relief, counterpiracy, near peer sea control, routine sea control, area of operations exercises) in terms of their duration, costs, success likelihood, risks, and specialized requirements. Losses in mission metrics captured the real-time harm of successful cyberattacks. Importantly, the prototype's mission selection section was deliberately left expandable to accommodate additional kinds of missions in the future.

Cyber operations were as deeply researched as navies, again, using only information available in open sources. Cyber was much more challenging, however. There is no open source intelligence database for cyber teams, skills, and operations.[21] We reviewed commercial job advertisements and consulted with a wide range of subject matter experts to construct a set or list of skills needed for defensive and offensive teams. The most common skill compositions extracted from commercial cyber defense sources were the threat assessment team, network security team, information security team, and incident response team. The typical commercial penetration team was used as a surrogate for an offensive team. These included a network penetration team, covert operations team, information exploitation team, and system exploitation team. Over time, our database for the game included considerable data on a wide array of cyber professional attributes, such as salary ranges per specialty, the correspondence between specialty skill levels and likely task success, and the cumulative effects of choosing one cyber team over another for a given task.

We roughly tied cyber operational tasks to the cyber kill chain.[22] For gamification and to replicate reality, performance attributes were distributed so that no team was equally successful at all tasks. Nor could a dominant strategy ensure a win every time.[23] Defensive tasks ranged from network hardening, auditing, and creating or restoring backups to reconfiguration and quarantine. Offensive tasks included gathering supply chain or network access data, configuring backdoors, changing database files, extracting operational data, analyzing systems dependencies, and taking targeted systems offline, among others.

Every task force was limited to three teams to induce players' strategic selection of teams for missions and anticipated needs, and to approximate limited talent in the real world. For example, at the outset of a game, a player picks three cyber teams to travel with their naval taskforce to a specific mission location. These three cyber teams could be all defensive, all offensive, or a mix, according to the player's expectation of what skills they would need to defend their naval task force or push back against adversaries during deployment at sea.

We added an interactive, third-party global map to the game to provide players with maritime-themed visuals to help hold their attention. We also provided an interactive tracker to the game dashboard, indicating fleet destination, the intended path, and current progress. As it happens, these decisions proved consequential for this prototype's development schedule. Embedding the map in game code slowed multiplayer gameplay interactions. Moreover, it proved difficult to keep the map synchronized with the game elements as more players were added.

A key goal was to have internal standardized and automated adjudication in game time. Recall that we wanted to avoid turn-based play or a reliance on non-standardized decision-making by human judges (i.e., white cells). Adjudication algorithms are key to virtual games. They define the relationships, calculations, win/lose conditions, and essential interactions. To avoid burying algorithmic decisions in the prototype game's code, we mirrored the algorithms in an

Excel spreadsheet for tracking and adjustments. This documentation made the underlying coding decisions more transparent. The spreadsheet also supported our future goal of developing a dashboard for easier adaptations to the game by nontechnical researchers or instructors, as well as for eventually making the game into a single-person game played against an AI opponent.

We playtested the game sporadically in the first year while collecting naval and cyber data and embedding it in code. Playtesting increased during the second year. Ironically, an unexpected downside of playtesting was that the players asked for improvements beyond the reach of our small development team. For example, we added military alliances and a chat system through which allied forces could share cyber defense resources. These additions increased the utility of the game for education and even research on operational cooperation. Yet while these additions functioned at a basic level, full functionality and an accompanying tutorial were still being built when game development had to cease.

By the fall of 2019, our team had created a working prototype, and we named the game *Cyber Maritime Common Defense*. It functioned as a two-person game for players on the same network. Each player was able to defend their fleet's network systems and to attack an adversary. In the opening scenario, a player is given a mission and an area of operations. They choose their task force command ship and three accompanying cyber teams.

For example, with a freedom of navigation mission in the Bering Strait, one player's task force sails the strait. While en route, that player engages their defensive team(s) in detecting and repulsing the other player's cyberattacks; they can also use their offensive cyber team(s), if they chose any, to disrupt their opponent. The main performance measures that contribute to winning are safe movement to the area of operation without critical ship systems being significantly hindered by the adversaries' cyberattacks. The visual measure of success is an "operational functionality bar," which changes from green to orange to red if the player's cyber defense efforts prove insufficient and their ships are losing network functionality over time. The "win condition" of any specific game is determined by the game host as either playing a set time, accomplishing a set number of mission transits on time, or remaining the last task force standing after all others are driven to zero functionality by cyberattacks.

Our funding ended in December 2019. For better or worse, this offered a pause for reflection. On the whole, the project developed a working prototype but suffered some significant constraints and missteps. While our most ambitious goals have not yet been reached, the experimental effort provided several useful lessons for others who may be considering such a project. It also built a good foundation of data, mechanics, and algorithmics needed to wargame cybered conflict in a virtual format.

PROCESS ANALYSIS AND LESSONS LEARNED

There remains no standard, best way to produce a virtual game, serious or not.[24] The gaming literature tends to agree that game designers need clear statements of rules, objectives, and strategies. They must use playtesting to ensure that the game is engaging, functions as intended, and meets the audience's needs. They must also expect their prototypes to fail repeatedly.[25]

A serious cybered conflict game would be challenging to build in any case.[26] Players must acquire specialized knowledge and facts and yet enjoy the iterative replay, complicating the coordination between the educators' or researchers' desires and the production processes of designers or coders who build the game.[27] An audience that is unfamiliar with games or has little time to play them—in our case, senior military officers—increased the game design and build complexity.

In evaluating our experience in hindsight, our project offers at least three key lessons for fellow travelers in virtual wargaming.

Balance Realism and Engaging Play

In the beginning, our objective was to build a wide variety of modular elements that could be tailored into different scenarios, involving different roles (ranging from states to corporations), for different players, across a range of different win conditions. As Mitgutsch and Alvarado would argue, the breadth of our desire inadvertently skewed our initial focus toward game mechanics involving content and aesthetics over any predetermined narrative or audience.[28] Our focus also encouraged a design process that emphasized menus of actions that could be used in any number of different stories. Over time, our layout of action options presented in intricate menus began to dominate the aesthetics as well.

Unintentionally, the fun of a real time game became subordinated to the work of building a game for players who were already assumed to be highly motived by prior interest in cyber actions, naval campaigns, and attack effects. For instance, we spent a great deal of time tweaking underlying algorithms to avoid giving any cyber defensive or offensive team superhero status, whereby using those teams would become unrealistically dominant. Our game slowly iterated away from a focus on player engagement and immersion to a game for designers, project teams, or senior leaders. If the intended players are not interested, however, the value and effectiveness of a serious game as a "transformational" tool for education and organizational change can be lost, as discussed in chapter 11.

Resource a Complete Team

Building something new while working within a government institution can be difficult, given restrictions ranging from budget to security requirements. Our funding only allowed a single hire, but a game designer is not a good substitute for a coder's skills.[29] When combined with the multiple objectives described above, the resulting code lacked streamlining and elegance.

This difficulty was further compounded by the need to use a particular authorized version of the Unity engine tied to a particular year. As the open source engine itself evolved, the project code had to stay with the more dated version. A better choice might have been a more light-weight open source engine that was more up to date. However, that was not a viable option for us.[30]

Start Simply with Vision

Building the third prototype as a multiplayer rather than a single-player game was a consequential decision. The multiplayer, multiscenario, modular game vision made it difficult for us to envision and then create simpler versions of the primary game mechanics using, say, paper or cards. Documentation and review can also be tedious and tempting to avoid if the game starts off being too complex. And the more time passes, the more production staff resist retrospectives and reboots.

Ironically, perhaps, our desire to make the game code transparent for non-specialists and easy to tailor, update, and upscale unwittingly reinforced the complexity of the coding challenge. Without highly skilled coding and professional documentation of the code base, the game became overly dense with loops and connection, which commensurately slowed the speed of resolution of gameplay. These are the risks associated with trying to build a multiobjective game.

MORE FOR THE FUTURE

At the end of the day, the lesson to learn from our experience is *not* to avoid virtual games but rather that the gameplay, design vision, and ultimate objective need to have a simpler caterpillar stage before the final butterfly. A better plan would have been to target one feature at a time and make that tangible in code and documents before allowing the understandable enthusiasm of coders, project teams, and playtesters to add more. That said, going through as many as ten failed prototypes before building a good game is normal. "Until you have two completely finished levels," Schell advises, "you are still figuring out the fundamental design of your game."[31] This advice is critical for any game, but doubly so for a serious, virtual game depicting the complex reality of cybered conflict.

As we continue to develop *Cyber Maritime Common Defense*, we expect to step back and refocus the game on a single audience with less fidelity, fewer options, a more streamlined user interface, and more fun. The game we envisioned in 2016 still does not exist. But the need remains.

Continued work on projects like this one is exceptionally worthwhile. Emerging technologies such as artificial intelligence and quantum computing are cyber's offspring. When widely distributed across military and civilian systems, they will extend the speed and reach of adversaries enough to extinguish the time for humans to pause and reflect when surprised. They will inherit whatever vulnerabilities that have not been resolved in the underlying cyber substrate on which they are developed and depend.

Defenders need to keep pace by constantly challenging themselves—and their increasingly autonomous and artificially intelligent systems—in games. If the wargame allows the potential future decision-makers or developers and users of these systems to prepare even partial but effective responses in advance of the real-life event, then the likelihood increases that the right choices will be made in real life, despite unanticipated events and limited time to respond. Virtual wargaming is essential to prepare cognitively for the complex and dangerous surprises coming in an emergent deeply digitized and conflict-prone world.

NOTES

1. Warwick Ashford, "Europe Faces Shortage of 350,000 Cyber Security Professionals by 2022," *Computer Weekly*, June 6, 2017, www.computerweekly.com/news/450420193/Europe-faces-shortage-of-350000-cyber-security-professionals-by-2022.
2. Dan Lohrmann, "Why Many Organizations Still Don't Understand Security," *Government Technology*, February 19, 2019, www.govtech.com/blogs/lohrmann-on-cybersecurity/why-many-organizations-still-dont-get-security.html.
3. Cybered conflict is the spectrum of conflict between states ranging from peace to war as enabled by the spread of the insecurely created global Internet. It has various names across main adversaries from the latest "information warfare" (United States), "hybrid warfare" (Russia), and "informatized warfare" (China), but the key is the spectrum largely occurs as whole societal system versus system contestation without military kinetic exchanges. See Chris C. Demchak, *Wars of Disruption and Resilience: Cybered Conflict, Power, and National Security* (Athens: University of Georgia Press, 2011).
4. We would like to thank Christopher Marques, whom we hired straight out of the New England Institute of Technology, for all his design and coding efforts for this game experiment. We could never afford to give him all the coding support needed, but he soldiered on determinedly and creatively to our great and enduring appreciation. We would also like to thank the long-standing support of professors

Alfred Turner and John Hanus of the US Naval War College for their constancy and kindness in participating in advisory reviews throughout progress of the game experiment.

5. Steve Ranger, "World's Biggest Cyber Wargame Features Battle Over Online Services and Industrial Control System—'Berylia' May Be a Fictional Battleground but the Issues Involved Are Perfectly Plausible," ZDNet.com, April 18, 2018, www.zdnet.com/article/worlds-biggest-cyber-wargame-features-battle-over-online-services-and-industrial-control-system/.

6. Sascha Dov, Andrew Dowse, and Hakan Gunneriusson, "Competition Short of War: How Russia's Hybrid and Grey-Zone Warfare Are a Blueprint for China's Global Power Ambitions," *Australian Journal of Defence and Strategic Studies* 1, no. 1 (2019).

7. The cybered conflict spectrum ranges from peace to war and is a system-versus-system contestation between states that involves the whole of society. Chris C. Demchak and Peter J. Dombrowski, "Rise of a Cybered Westphalian Age," *Strategic Studies Quarterly* 5, no. 1 (March 1, 2011).

8. These are the unprecedented five offense advantages made available by the global spread of a very insecurely coded underlying cyber substrate. For a more extensive explanation, see Chris C. Demchak, "Cybered Conflict, Hybrid War, and Informatization Wars," in *Handbook of International Cybersecurity*, edited by Eneken Tikk and Mika Kerttunen (New York: Routledge, 2020).

9. Rain Ottis, "Light Weight Tabletop Exercise for Cybersecurity Education," *Journal of Homeland Security and Emergency Management* 11, no. 4 (2014).

10. While many card games had by then been reproduced online with computer opponents, that level of artificial intelligence is less the traditional card game than a more virtual game. See, e.g., the review work of Maria Kordaki and Anthi Gousiou, "Computer Card Games in Computer Science Education: A 10-Year Review," *Educational Technology & Society* 19, no. 4 (2016).

11. The online digital card game for education was at the time also not considered sufficiently mature for serious education. See A. Gousiou and M. Kordaki, "Digital Card Games in Education: A Ten Year Systematic Review," *Computers & Education* 109 (June 2017).

12. Jesse Schell, *The Art of Game Design: A Book of Lenses* (London: Taylor & Francis, 2008).

13. Margaret L. Loper, Charles Turnitsa, and A. Tolk, "History of Combat Modeling and Distributed Simulation," in *Engineering Principles of Combat Modeling and Distributed Simulation*, edited by A. Tolk (Hoboken, NJ: Wiley, 2012).

14. David C. Lane, "On a Resurgence of Management Simulations and Games," *Journal of the Operational Research Society* 46, no. 5 (1995).

15. Anna K Preuß, "The Learning Process in Live-Action Simulation Games: The Impact of Personality, Motivation, Immersion, and Flow on Learning Outcome in a Simulation Game," *Simulation & Gaming* 52, no. 6 (2021).

16. The US Department of Defense has identified three kinds of simulations since at least the 1990s, namely, live, virtual, and constructed. But limitations on computing power, coding, data, and skills of the designers resulted in the resulting virtual aspect of Defense exercises being more simulations than replayable games. See Lane, "On a Resurgence."

17. David R. Krathwohl, "A Revision of Bloom's Taxonomy: An Overview," *Theory into Practice* 41, no. 4 (2002).

18. Kuan-Ta Chen, Chun-Ying Huang, and Cheng-Hsin Hsu, "Cloud Gaming Onward: Research Opportunities and Outlook," paper presented at 2014 IEEE International Conference on Multimedia and Expo Workshops, 2014.

19. Tony Manninen, "Rich Interaction in the Context of Networked Virtual Environments: Experiences Gained from the Multi-Player Games Domain," in *People and Computers XV: Interaction without Frontiers*, edited by Ann Blandford, Jean Vanderdonckt, and Phil Gray (New York: Springer, 2001).

20. "Unity is a game engine and integrated development environment (IDE) for creating interactive media, typically video games." John K Haas, *WPI Digital Commons: A History of the Unity Game Engine* (Worcester: Worcester Polytechnic Institute, 2014), https://digitalcommons.wpi.edu/iqp-all/3207.

21. See "Janes National Security Online Database," www.janes.com/.

22. There are various versions of this kill chain, some with seven or eight steps, and others with over ten. The best known is the Lockheed Martin seven-step version: reconnaissance, weaponization, delivery, installation, command and control, and actions on objectives. See "Cyber Kill Chain," Lockheed Martin Corporation, no date, www.lockheedmartin.com/en-us/capabilities/cyber/cyber-kill-chain.html.

23. "When choices are offered to a player, but one of them is clearly better than the rest, this is called a dominant strategy. Once a dominant strategy is discovered, the game is no longer fun." Schell, *Art of Game Design*.

24. Annakaisa Kultima, "Game Design Research," paper presented at Proceedings of the 19th International Academic Mindtrek Conference, 2015.

25. Schell, *Art of Game Design*.

26. Konstantin Mitgutsch and Narda Alvarado, "Purposeful by Design? A Serious Game Design Assessment Framework," in *Proceedings of the International Conference on the Foundations of Digital Games (FDG '12)* (New York: ACM, 2012), 121–28.

27. Bertrand Marne et al., "The Six Facets of Serious Game Design: A Methodology Enhanced by Our Design Pattern Library," paper presented at European conference on technology enhanced learning, 2012.

28. Mitgutsch and Alvarado, "Purposeful by Design?"

29. Richard Carrillo, *Lead Game Designer: Visionary, Empath, Salesman, Analyst* (Toronto: Ubisoft, 2015), www.gdcvault.com/play/1021923/Designing-Your-Design.

30. "The major components of a typical game engine would be the rendering engine to represent the virtual game world onto the screen, the audio engine to produce music and sound effects, the physics engine to apply natural motion effects between objects, the event system to handle various events that occur on the screen and other devices, the game logic system, and others. Additionally, the network system, the database system and the animation system can be additionally integrated into the game engine." Haechan Park and Nakhoon Baek, "Developing an Open-Source Lightweight Game Engine with DNN Support," *Electronics* 9, no. 9 (2020).

31. Schell, *Art of Game Design*.

13

BREATHING LIFE INTO MILITARY DOCTRINE THROUGH CYBER GAMEPLAY

BENJAMIN C. LEITZEL

In the profession of arms, military officers must come to terms with complex technical knowledge about cyberspace under conditions of competition and conflict. They must also make sense of other elements in military doctrine. I argue that cyber wargaming can support learning and retention of even the seemingly dull and complicated content involved. This is true for civilian education about cyber policy (e.g., chapters 8 and 9) and workplace training about cyber incident response (e.g., chapters 10 and 11). Although the military has different challenges, the value of wargames for educating the officer corps remains (see also chapter 12).

For better or worse, military doctrine matters—in cyberspace, as elsewhere. Doctrine is how the armed services define the terminology they use and outline the fundamental principles of action for leaders, planners, and operators.[1] Doctrine provides commanders with a common framework for guiding and communicating their action in the complex, interactive, and even lethal environments where the armed services operate.

Understanding doctrine does not mean mindlessly following fixed or rigid rules. Competition and armed conflict are far too complex and interactive to master with preset rulemaking or checklists. Instead, effective doctrine is more akin to the "rules of the road" in driving. While specific traffic laws vary between national and local jurisdictions, good drivers must understand the fundamental principles to enhance safety for all (e.g., obeying speed limits, maintaining a safe following distance, signaling intentions, and heeding traffic signs).[2] Like the rules of the road, military doctrine provides "fundamental principles that guide the employment of US military forces toward a common objective."[3]

Simultaneously, knowing the fundamental rules of the road is also key for knowing when drivers must step outside of the guidelines, such as executing an unauthorized lane change to avoid a collision. Military leaders are likewise reminded to align their actions to doctrine so that they can decide when exceptions are necessary. The point is that strict adherence in either driving or observing doctrine is fundamentally wrong. Instead, knowing the principles allows for drivers and leadership to use judgement before deciding.

To this end, the Army War College's mission is to "educate and develop leaders for service at the strategic level," as well as to "think critically and strategically in applying joint warfighting principles and concepts to joint operations."[4] These are high standards, requiring a depth of understanding and critical thinking far beyond a casual acquaintance with the material. In keeping with this mission, students at the US Army War College are required to understand the military cyberspace missions and activities described in Joint Publication 3-12, *Cyberspace Operations*. Simply put, this doctrinal publication "establishes a framework for the employment of cyberspace forces and capabilities."[5] Students typically start to learn about this doctrine in the classroom through traditional means, such as assigned readings and seminar discussions. Through these readings and discussions, students are expected to understand and assess the operational and strategic implications of military operations in cyberspace.

This is easier said than done. One challenge with effectively teaching and learning this content is that military doctrine is important but also complicated and dense. It can be extremely boring to study. Like many rulebooks, doctrinal publications often include hundreds of pages of mundane text. Furthermore, rote memorization is not sufficient. Military leaders must understand doctrinal terminology and principles to effectively enable their decisions and actions outside the classroom.

Cyberspace compounds the challenge of teaching and learning military doctrine. As suggested throughout this book, cyber operations stand to upend much of what the Army thinks about geography, weapon systems, and combat. It is all too easy to assume that our students are so-called digital natives who grew up using network information technology and therefore understand how cyberspace works. Not so. Just because students use cyberspace does not mean that they understand its underlying mechanisms, let alone how militaries can fight wars in and through it.

Once you get past the glossy veneer of graphical user interfaces to explore the how, why, and what of military cyber, the content can become so technical that students have difficulty connecting with it, let alone understanding and retaining the fundamental principles. Many cyber operations are also classified. Simply trying to discuss real-world examples can sometimes vault an entire conversation into secret, top secret, and even higher levels of classification. As a result, unclassified discussions must often remain relatively abstract. Some students "hit a wall," unable to absorb, apply, or even engage with this material. Personally, this experience reminded me of an unfortunate high school teacher in the movie *Ferris Bueller's Day Off*, who struggles to engage his class on the important but dull topic of the Hawley-Smoot Tariff Act. His pleas for student input are met with resounding silence: "Anyone know what this is? Class? Anyone? Anyone? Anyone seen this before?"[6]

To make these lessons stick, we need to breathe life into military doctrine. For myself and my colleagues, we found that rote dictation and unclassified reading fell far short of the learning outcomes demanded by our students, the Army, and the Joint Force. We therefore decided to design a hands-on cyber wargame to put our students—mid-level and senior military officers—in fictional but realistic scenarios to help them experience, learn, and apply military doctrine in an unclassified classroom.

The value of such experiential learning is, admittedly, taken for granted in professional military education, since servicemembers are expected to train and drill throughout their careers. However, robust analysis of the pedagogical value of wargaming in professional military education is limited. Experiential learning through game-based learning and gamification is a vibrant subfield of pedagogy in other areas of the social sciences, humanities, and business management. This interdisciplinary research typically finds that student engagement and comprehension increase when gameplay is coupled with assigned reading and course discussion.[7] As described in chapter 8, the notion of "play" is also crucial for educational wargaming, bringing students into a "lived," or emotionally and intellectually engaged, learning cycle.[8] Despite the apparent overlap between wargaming and pedagogy, as well as their relevance for professional military education, these bodies of literature often run parallel to, rather than mutually inform, one another.

To help bridge this divide, advance our teaching practice, and improve student learning, the faculty of the US Army War College's Center for Strategic Leadership teamed up to develop a simple and unclassified cyber wargame in 2019. Our goal was to introduce students to joint military doctrine for cyberspace operations, missions, and actions. By merging the printed word (doctrine) with experiential learning (the game), we designed a tool that would allow current and future leaders to understand and critically analyze the technicalities and abstractions of military cyberspace more effectively than more traditional teaching and learning methods.

GAME DESIGN

The process of creating a new cyber wargame entailed several months of development and dozens of playtests before deployment with students in the classroom. We began with a few overarching requirements. The game needed to be based on current military doctrine; to be simple enough to employ with limited facilitation; to fit into a three-hour block of instruction that included an introduction, overview, and summary; and to be engaging for players.[9] More specifically, the game needed to help students understand cyberspace missions and actions described in Joint Publication 3-12, *Cyberspace Operations*; analyze cyber effects on friendly and adversary forces; and evaluate the resulting changes in friendly and adversary capabilities, requirements, and vulnerabilities across the conflict continuum.

To emphasize the link between cyber effects and the commander's intent to accomplish mission objectives, our design team incorporated "Joint Functions" into the game. Joint Functions describe the combination of expertise across the US military—the Army, Navy, Air Force, Marine Corps, Space Force, and Coast Guard—rather than each service's single or specific capabilities.[10] Military doctrine for joint operations defines these Joint Functions as "related capabilities and activities grouped together to help Joint Force Commanders integrate, synchronize, and direct joint operations."[11] These functions can be viewed as the essential elements of nearly all military operations. Commanders develop plans that group functionally related capabilities and activities into the seven joint functions: command and control, information, intelligence, fires, movement and maneuver, protection, and sustainment.[12] By focusing on Joint Functions at the operational level of war (rather than the highly classified tactical level), we avoided the classification issues that accompany tactics, techniques, and procedures, and elevated the issues of greatest concern for senior military leaders.

We also decided that gameplay should span the conflict continuum. The conflict continuum matters because military operations are conducted with the oversight of civilian leadership, and as potential conflict moves from peace to outright war, the political realities and authorities change. According to joint doctrine, this continuum runs from military engagement, security cooperation, and deterrence to crisis response, limited contingency operations, and large-scale combat operations.[13] Commanders must understand that, in the United States, civilian leaders must approve military operations, and that civilian leadership will place different restrictions on these operations at different times. For deterrence, for example, civilian leaders may allow limited cyberspace operation to prevent aggression, but they might restrict attacks against an adversary's critical infrastructure that may be otherwise entertained during large-scale combat operations.

With these objectives in mind, our design team explored options to accurately replicate cyberspace effects. Joint Publication 3-12, *Cyberspace Operations*, identifies three missions and four actions (see figure 13.1). In order to successfully execute cyberspace operations, commanders must integrate and synchronize cyberspace missions: first, Department of Defense Information Network (DODIN) Operations; second, Defensive Cyberspace Operations (DCO); and third, Offensive Cyberspace Operations (OCO).[14] Because military doctrine is complicated, like the reality it attempts to address, DCO is further

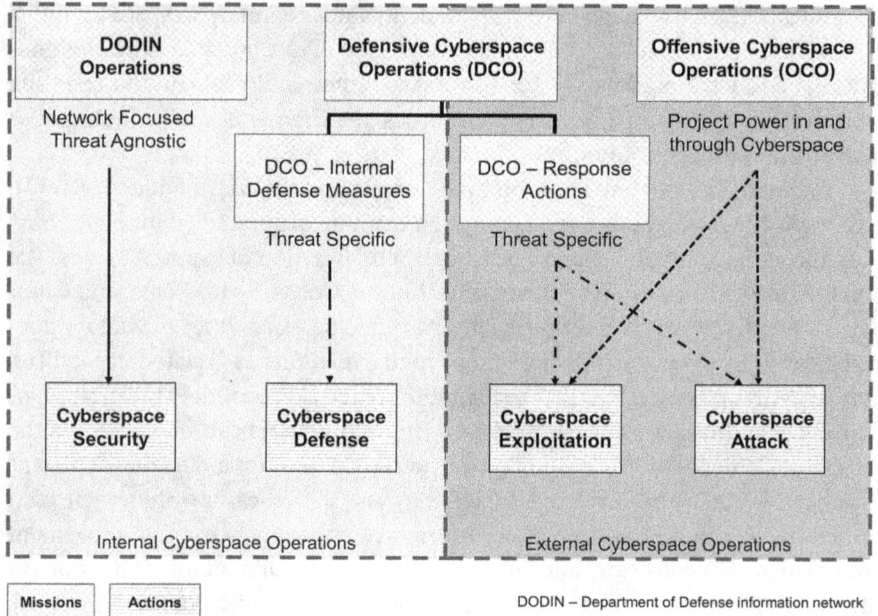

FIGURE 13.1. Cyberspace Missions and Actions. *Source:* US Joint Chiefs of Staff, *Cyberspace Operations*, Joint Publication 3-12 (Washington, DC: US Joint Chiefs of Staff, 2018), figure II-1, modified for clarity.

divided into Internal Defensive Measures (DCO-IDM) and Response Actions (DCO-RA). Cyberspace forces in the military act with the aim of creating specific effects in cyberspace. The four basic actions are security, defense, exploitation, and attack.[15]

PLAYTESTING

Once our team agreed on a basic format, cyberspace faculty and wargaming personnel playtested the game to ensure that it provided incentives for students to explore cyberspace missions and actions aligned with joint doctrine. During early playtesting, we observed that some players ignored realism to win the game by pursuing tactics that would not be possible in real life. This is the risk of negative learning, and we needed to adjust game mechanics to ensure that students were not learning the wrong lessons to simply prevail under artificial conditions. We invited gaming experts to do their best to win the game while disregarding doctrine and cyberspace fundamentals, and then we fixed the unrealistic quirks that they revealed. This round of testing resulted in significant changes.

For example, under the initial rules, players could accomplish cyberspace attacks without first conducting cyberspace exploitation. This option conflicts with doctrine, which states that "cyberspace exploitation actions are taken as part of an OCO or DCO-RA mission."[16] It is also in conflict with cyber operations in real life, which often require reconnaissance of a target before attack. Collecting and analyzing observations such as these through feedback from faculty and staff improved the game.

CYBER AS A GAME BOARD

Given the learning objectives and requirements for a simple, unclassified format, Major Krisjand Rothweiler proposed a board game based on Hasbro's classic naval combat game, *Battleship*®. Instead of attacking an opponent's ships, Rothweiler designed the game board so that players could use cyberspace actions to support or defend friendly joint functions while disabling the adversary's functions across the conflict continuum. Representing the environment in this way, players play the role of cyber planners supporting a larger military operation.

The game board allows players to defend friendly joint functions and attack the joint functions of their adversary. A keen observer of US military doctrine might point out that the information joint function appears be missing. However, the game is played entirely within what doctrine would refer to as the "information environment" (figure 13.2).[17] (On a more practical note, the design team also decided to include only six joint functions due to the limited availability of seven-sided dice.)

Players have limited resources to purchase tokens that represent different defensive and offensive missions and actions (see figures 13.3 and 13.4). We designed the game to advantage the offense and implemented rules to increase the likelihood of a successful cyberspace attack, based on faculty input and the rough—although contested—consensus in the literature about this domain. For example, Ben Buchanan, author of the book *The Cybersecurity Dilemma*, argues that "there's nearly universal acknowledgement that offense has the advantage."[18] We made cyberspace offensive tokens more expensive but also gave them a greater chance of success. It is important to note that while we biased our game in favor of offense, similar adjustments can make cyber defense equal or stronger based on different assumptions about the domain. Thus, if the prevailing assumption changes, the game can be easily modified.

Players can purchase and use six different kinds of tokens in the game, as illustrated in figure 13.3.

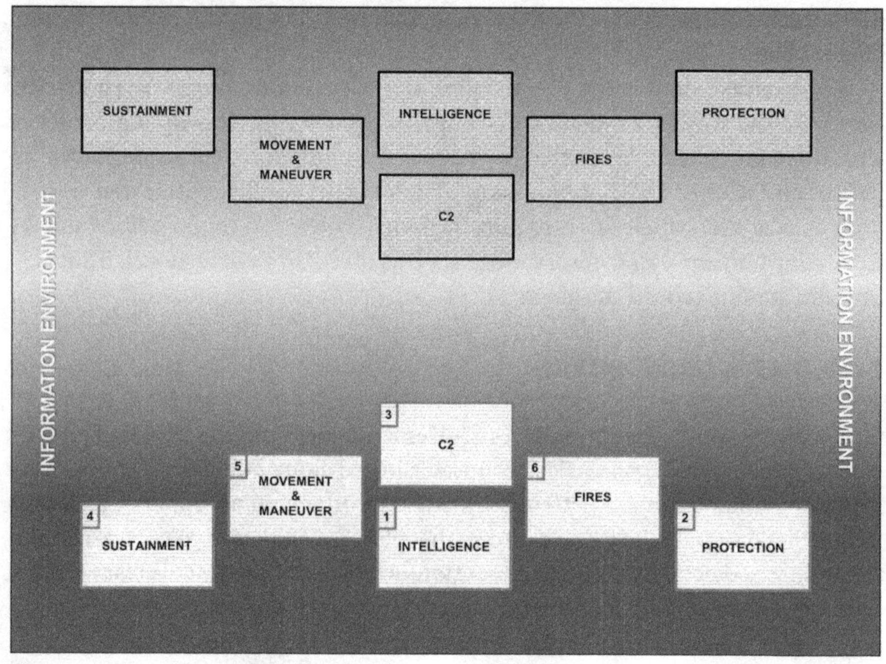

FIGURE 13.2. Cyber Game Board

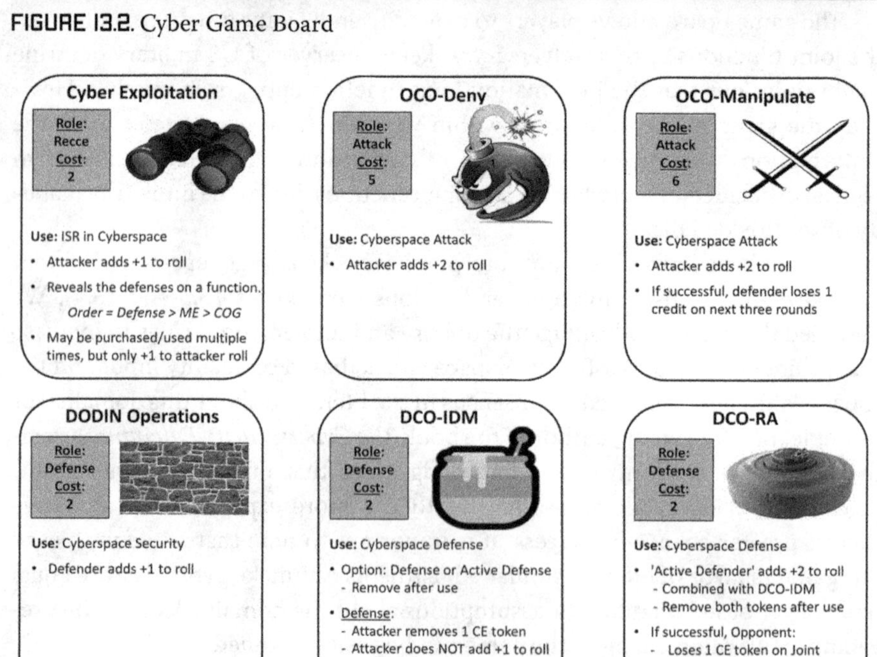

FIGURE 13.3. Cyber Game Tokens

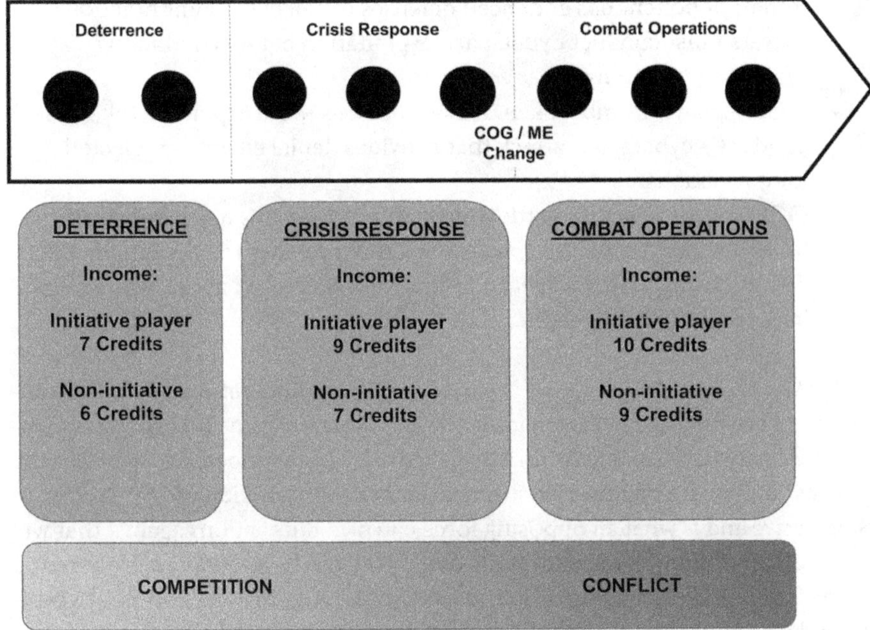

FIGURE 13.4. Cyber Game Phases and Income

These tokens are as follows:

- Department of Defense Information Network (DODIN) Operations: "Firewall" token. Players can position DODIN Operations on selected joint functions in order to enhance cybersecurity. These tokens are relatively inexpensive and add 1 point to a two-die roll. DODIN Operations tokens remain effective throughout the game.
- Defensive Cyberspace Operations–Internal Defensive Measures (DCO-IDM): "Honey Pot" token. Players can choose this token to identify adversaries on the selected joint function networks. Or they can combine it with a DCO-RA token to conduct active defense and strike back at adversary cyber teams. Players remove these tokens after use and need to repurchase them for additional defense and active defense actions.
- DCO Response Actions (DCO-RA): "Land Mine" token. When combined with a DCO-IDM, players conduct active defense to remove adversary cyber teams from the defended network. Players remove these tokens after use and need to repurchase them for additional active defense actions.
- Cyberspace Exploitation: "Binoculars" token. Players use this token to conduct intelligence on adversary networks in order to determine

if their opponent has enhanced defenses on specified joint functions. Players must conduct cyberspace exploitation before conducting offensive operations.

- OCO–Deny: "Bomb" token. Players use this more expensive token to conduct a cyberspace attack that provides denial effects on selected joint functions.
- OCO–Manipulate: "Swords" token. Players use this most expensive token to conduct a cyberspace attack that provides manipulation effects. The attacking player will earn points for a successful attack while reducing credits to the defender.

Finally, the income or payoff is varied to reflect different phases in the continuum of conflict, as illustrated in figure 13.4. Our wargame incorporates Centers of Gravity (COGs) and Main Efforts (MEs). Military doctrine describes the COG as "the source of power or strength that enables a military force to achieve its objective and is what an opposing force can orient its actions against that will lead to enemy failure."[19] Military staffs determine the COG for both friendly and enemy forces and then design plans to protect the friendly COG while attempting to defeat the enemy COG. The commander directs the ME to "concentrate capabilities or prioritize efforts to achieve specific objectives."[20] For example, during combat operations the ME concentrates kinetic fires on enemy forces, while the friendly COG is command and control (C2). Without effective C2, the commander would be unable to synchronize operations and the mission could fail. During the course of conflicts, COGs "may be transitory in nature" and "can shift over time or between phases."[21] The same is true of the ME.[22] We also gave proactive teams an advantage over reactive adversaries; therefore, the game provides more income to the "initiative player."

GAMEPLAY

Our wargame provides students with an opportunity to explore the impact of cyberspace effects on military mission objectives in a competitive yet fun environment. The game consists of three phases (deterrence, crisis response, and combat operation), with eight rounds therein that span the conflict continuum. After two rounds of play in the deterrence phase, players increase resources (credits) as they compete during three rounds of the crisis phase and then change their center of gravity, main effort, and resources for the final three rounds of the combat operations phase. Players can track the current phase and round using the graphic at the top of figure 13.4.

Each round consists of six steps. First, at the beginning of the game, players draw tokens to conduct cyberspace exploitation and defense (1 CE, 1 DODIN, and 1 DCO-IDM token). Next, players roll two dice to determine which joint functions are their COG and ME (associated with number in the upper left corner of joint function box, as shown in figure 13.2). The first die roll corresponds to the COG. The second die roll corresponds to the ME (reroll if the same as COG). During playtesting, the design team discovered that it was a significant disadvantage to have a single joint function as both the COG and ME. At the end of the crisis response phase, players roll to determine a new COG and ME.

Third, to determine which player has the initiative, both players roll one die and the player rolling the highest number starts the round with the initiative. Fourth, players collect income for the round (see figure 13.4). Players may use credits to purchase additional tokens or save credits for subsequent rounds. Moving defenses from one function to another also costs 1 credit.

Fifth, players make their moves by placing their chosen tokens on the game board. The order of play starts with passive defense: Both players simultaneously place DODIN and DCO-IDM tokens on their Joint Functions. Next is exploitation: Both players simultaneously place all CE tokens on their opponent's Joint Functions. The player with initiative conducts exploitation activities followed by opponent. Each player conducting CE will ask their opponent if they have cyberspace defenses on the Joint Function in question, or if the Joint Function is a COG or ME. Attack is next: The player with initiative conducts all their desired OCO actions, which are followed by their opponent. Then comes active defense: The noninitiative player conducts all desired DCO-RA actions, followed by initiative player. Sixth and last, each subsequent round starts at step 3 (unless the crisis response phase is complete, in which case they start the next round at step 2).

The player with the most points at the end of the game wins. Players accumulate points by conducting successful cyberspace attacks, defenses, and active defensives (see figure 13.5). Based on playtesting and feedback from subject matter experts, the design team adjusted points and credits to ensure that students could see the benefit of conducting each cyberspace mission and action:

- Cyberspace Attack (OCO) on COG: If successful, the attacker accumulates 3 points. If unsuccessful, the defender accumulates 1 point and 1 credit.
- Cyberspace Attack (OCO) on the ME: If successful, the attacker accumulates 2 points. If unsuccessful, the defender accumulates 1 point.
- Cyberspace Attack (OCO) on any other Joint Function: If successful, the attacker accumulates 1 point. If unsuccessful, the defender accumulates 1 point.
- Active Defense (DCO-RA, in combination with DCO-IDM): If successful, the defender accumulates 1 point and the opponent loses 2 credits. If unsuccessful, the opponent earns 1 credit.

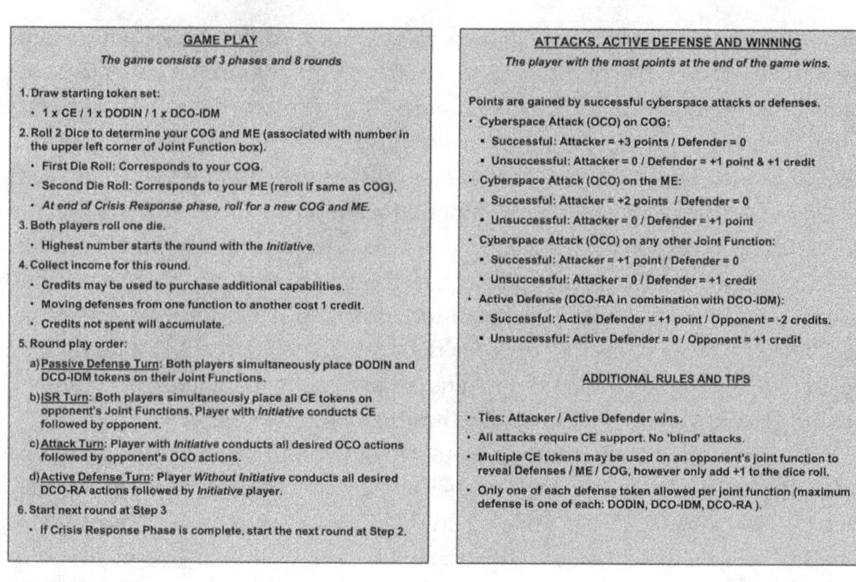

FIGURE 13.5. Cyber Game Instructions and Scoring

Students typically complete all rounds in 90 minutes. However, faculty can shorten the game once they determine that learning objectives have been accomplished.

OBSERVATIONS

After playing this game, our faculty observed that students were more interested and engaged with cyberspace doctrine. Incorporating the game into the classroom accomplished the desired learning objectives. Since the game has been included in cyberspace elective courses at the Army War College, student understanding and retention of this doctrinal material has improved, and faculty members have been better able to weave these concepts into later lessons that have explored more in-depth and complex issues involving cyberspace operations.

Student teams also employed creative tactics through the games, including deception. For instance, some players would place additional cyberspace defenses on less important joint functions. When the adversary team probed these joint functions, they would find extra defenses, assume that these functions were critical, and attack the lower-priority and lower-payoff targets. In addition, we observed that after playing a few rounds, students would often change tactics to reflect their deeper understanding of the actions and missions described by doctrine.

The game appears to help students grasp the subtle differences between DODIN operations, which are conducted in advance of any specific threat activity, and DCO, which defend against specific and active threats.[23] Similarly, gameplay helped students distinguish between DCO-RA—namely, defensive actions taken "external to the defended network or portion of cyberspace"—and OCO missions "intended to project power in and through foreign cyberspace."[24] As noted above, the game also reinforced joint doctrine by requiring students to initiate cyberspace exploitation actions to successfully accomplish OCO or DCO-RA missions.[25]

In addition, the game allowed students to analyze and evaluate changes in friendly and adversary critical capabilities, requirements, and vulnerabilities across the conflict continuum. While engaged in the current fight, students had to look ahead and prepare for each new phase as the game progressed from deterrence to crisis response and finally conflict. It was clear to the cyber faculty team that students' understanding of both cyberspace and other doctrinal concepts improved as a result of playing the game. We noted that students demonstrated more in-depth knowledge of cyberspace operations during seminar discussions and better incorporated fundamentals in their end-of-course papers on the critical analysis of strategic cyberspace issues.

Faculty also noted that, unlike the preparation of rote memorization, the game proved to be more easily adapted to changes in case studies (i.e., the context in which gameplay is framed), and in emphasis (i.e., defensive operations and offensive operations). Faculty members found it easier to update and challenge students as they modified the game to meet their objectives.

One of the most notable variations came from a faculty team that used this cyber wargame to support its end-of-course joint planning and execution exercise. It chose to conduct this exercise without the game board or tokens while still using a similar methodology to enhance the scenario. The instructor identified a few students from the larger group to be red and blue cyberspace planners. During each phase of the exercise, the students conducted offensive and defensive cyberspace operations in accordance with their commander's intent and objectives. As in the board game, students identified joint functions to defend and attack and rolled dice to determine success or failure. The lead faculty member adjudicated these results and provided updates on each event. For example, the red team conducted a successful cyberspace attack on blue's intelligence joint function. The faculty member chose to omit the location and movement of the adversary's naval forces during the next scenario update as if intelligence was blinded, thereby demonstrating cyberspace effects as part of the student team's broader plan.

CONCLUSION

This cyber board game exceeded our initial expectations. In summary, it is a simple and unclassified game that introduces students to cyberspace missions and actions. The game allows students to plan and execute cyberspace missions and actions to produce effects that accomplish a commander's intent and objectives. It also incorporates Joint Functions and spans the conflict continuum. After they play, students can easily reflect on this learning experience to address key elements of doctrine when exploring complex issues during the subsequent lessons of each course. Teachers can likewise refer back to aspects of the game to help students ground abstract concepts or technical content in concrete and comprehensible experience. It is a win-win, for teachers and students alike.

Of course, there is always room for improvement. For example, some subject matter experts and student players questioned the success criteria for cyberspace attacks. In order to provide a more sophisticated understanding of offensive techniques, tactics, and procedures, the game would need to incorporate more detailed descriptions of the cyber vulnerabilities of military systems. In some cases, doing so would also require a secure facility and limit play to students with security clearances.

Furthermore, the game addresses joint military functions, but it omits related and potentially interdependent effects of strategic attacks against critical infrastructure in the civilian sector, such as the electrical grid discussed in chapter 10. And while our board game uses token cards and dice, some observers noted that the game could be converted to a computer-based, online version, as described in chapter 12. Others argued that the face-to-face and physical components of our game enhanced competition, fun, and experiential learning.

Potential modifications aside, the takeaways from this cyber wargame should not be overlooked. Doctrine is an incredibly important element of professional military education. This game proved helpful, especially for officers who are not steeped in cyber expertise. Even for officers who are familiar with the domain, this type of game can encourage creative and critical thinking about the strengths and weaknesses of current doctrine—doctrine they may one day help to update and improve. Gameplay opened the discussion space for students to interrogate their own decision-making, rather than simply taking the requisite decisions and outcomes as self-evident or set in stone. Finally, this game managed to address cyber missions and actions without raising the thorny issue of classification. Although the pedagogical value of wargaming in professional military education warrants further examination (cyber or otherwise), our experience with this form of experiential learning suggests considerable promise for outcomes-based education.

NOTES

1. Among other scholarly studies on the origins and potential pathologies of military doctrine, see Barry R. Posen, *The Sources of Military Doctrine* (Ithaca, NY: Cornell University Press, 1984); Jack Snyder, *The Ideology of the Offensive: Military Decision Making and the Disaster of 1914* (Ithaca, NY: Cornell University Press, 1991); and Elizabeth Kier, *Imagining War: French and British Military Doctrine between the Wars* (Princeton, NJ: Princeton University Press, 1997).

2. "Rules of the Road: 11 Essential US Traffic Laws a Driver Must Know," ePermitTest, October 7, 2020, www.epermittest.com/drivers-education/basic-driving-rules.

3. US Joint Chiefs of Staff, *Doctrine for the Armed Forces of the United States*, Joint Publication 1 (Washington, DC: US Joint Chiefs of Staff, 2013; updated with change 1, 2017), IX.

4. US Joint Chiefs of Staff, *Officer Professional Military Education Policy*, Chairman of the Joint Chiefs of Staff Instruction 1800.01F (Washington, DC: US Joint Chiefs of Staff, 2020), A-3, A-B-8.

5. US Joint Chiefs of Staff, *Cyberspace Operations*, Joint Publication 3-12 (Washington, DC: US Joint Chiefs of Staff, 2018), VII.

6. *Ferris Bueller's Day Off*, directed by John Hughes (1986; Los Angeles: Paramount Pictures). Also see *Ferris Bueller's Day Off*, character: Ben Stein: economics teacher, IMDB, 1986, www.im,db.com/title/tt0091042/characters/nm0825401.

7. See Mert Kartal, "Facilitating Deep Learning and Professional Skills Attainment in the Classroom: The Value of a Model United Nations Course," *Journal of Political Science Education*, 2020, 1–21; and Petra Hendrickson, "Effect of Active Learning Techniques on Student Excitement, Interest, and Self-Efficacy," *Journal of Political Science Education*, 2019, 1–15.

8. E.g., players need to buy into the scenario and flow of events to engage effectively; see Peter P. Perla, and E. D. McGrady. "Why Wargaming Works," *Naval War College Review* 64, no. 3 (2011): 111–30.

9. Credit should be given here to Major Krisjand Rothweiler, who laid out a thorough plan for game development to included interviews with faculty regarding senior-leader cyberspace education challenges and observation of several cyberspace seminars lessons. Major Rothweiler noted that faculty members insisted on the proper use of doctrinal terminology and consistently emphasized the importance of using cyberspace effects to protect friendly critical capabilities and requirements while attacking adversaries' vulnerabilities in order to accomplish the commander's objectives.

10. Thomas Crosbie, "Getting the Joint Functions Right," *Joint Force Quarterly* 94, no. 3 (2019): 108–12.

11. US Joint Chiefs of Staff, *Joint Operations*, Joint Publication 3-0 (Washington, DC: US Joint Chiefs of Staff, 2017; updated with change 1, 2018), III-1.

12. US Joint Chiefs of Staff, XIII–XV.

13. US Joint Chiefs of Staff, V-4–V-5.

14. US Joint Chiefs of Staff, Joint Publication 3-12, II-1.

15. US Joint Chiefs of Staff, II-5.

16. US Joint Chiefs of Staff, II-6.

17. Curtis M. Scaparrotti, *Information Operations*, Joint Publication 3-13 (Washington, DC: US Joint Chiefs of Staff, 2012).

18. See Alyza Sebenius, "Writing the Rules of Cyberwar," *Atlantic*, June 28, 2017, www
.theatlantic.com/international/archive/2017/06/cyberattack-russia-ukraine-hack
/531957/. Also see Ben Buchanan, *The Cybersecurity Dilemma: Hacking, Trust, and Fear between Nations* (Oxford: Oxford University Press, 2016). Critics of this conclusion include Jon R. Lindsay, "Stuxnet and the Limits of Cyber Warfare," *Security Studies* 22, no. 3 (2013); and Rebecca Slayton, "What Is the Cyber Offense-Defense Balance?" *International Security* 41, no. 3 (2016). Although the number of failed attacks is unknown, we decided to give offense the advantage in our cyber wargame given the large number of seemingly successful exploits and attacks, such as the Office of Personnel Management breach.

19. US Joint Chiefs of Staff, *Joint Planning*, Joint Publication 5-0 (Washington, DC: US Joint Chiefs of Staff, 2020), IV-22.

20. US Joint Chiefs of Staff, III-66.

21. US Joint Chiefs of Staff, IV-24 (figure IV-7).

22. US Joint Chiefs of Staff, III-67.

23. US Joint Chiefs of Staff, Joint Publication 3-12, II-2.

24. US Joint Chiefs of Staff, II-4 and II-5.

25. US Joint Chiefs of Staff, II-6.

14

MATRIX GAME METHODOLOGIES FOR STRATEGIC CYBER AND INFORMATION WARFARE

HYONG LEE AND JAMES DEMUTH

As policymakers' and military leaders' interest in cyberwarfare grows, wargame designers are increasingly asked to, as the joke goes, "sprinkle some cyber" onto their next game. This is easier said than done. For example, many of the intrusions and attacks that are often characterized as cyberwarfare are also characterized by anonymity.[1] This anonymity creates asymmetries in information environments, as described in chapter 5. Moreover, cyber capabilities are not discrete units, like brigades or fighter squadrons, and the terrain and time frames in which they operate differ from those for more conventional, kinetic capabilities, such as tanks and missiles. The basic challenge for game designers is how to effectively incorporate the unique features of cyber and information warfare into wargames, just as the basic challenge for players is to understand their implications.

In this chapter, we discuss our attempt to better represent cyber and information warfare in a series of strategic-level, matrix-game-based exercises called *Pinnacle Protagonist*. In doing so, we examine the challenges we identified with our original wargame design and how we attempted to address them. Our experience offers two broad lessons. First, changes to the flow of the game can provide opportunities for anonymous player action, even in principally open information wargames. Second, limiting players' opportunities to act can create certain realistic resource constraints in strategic-level games, even when specific resources or assets are not represented. Although some of our observations are specific to matrix games and our particular use case, we hope that our experience provides other designers with a useful example of how to approach the intimidating task of integrating cyber and information warfare into games at the strategic level.

MATRIX WARGAMES

Pinnacle Protagonist was designed as a three-day, end-of-year capstone exercise at the National Defense University's College of Information and Cyberspace.

We conducted this capstone for about fifteen students a year from 2017 to 2019. Our students were primarily midcareer US military officers and civilians. The objective of this capstone exercise was to provide faculty with an opportunity to observe and evaluate students' ability to apply the cyber and information warfare curriculum they learned throughout the academic year. For students, the exercise provided an environment where they could demonstrate the extent to which they had achieved the joint learning objectives outlined in official policies for professional military education.[2]

Given the diverse, abstract, and strategic nature of the curriculum, we decided to structure *Pinnacle Protagonist* around a series of matrix games. Matrix games are a particular style of tabletop, role-playing wargame. They use structured argumentation between players to facilitate collaborative adjudication and narrative building, in contrast to more rigid approaches to adjudication based on strict and predetermined rules.[3] The scenario centered on an escalating security crisis between the Baltic states and Russia, which entangled the United States, Western Europe, and the Nordic nations.

In a typical matrix game, players are given a role and take turns attempting to advance their interests in the wargame scenario. During their turn, players describe the action they are taking, its intended effect, and why they think it will succeed. To adjudicate the outcome, other players make arguments as to why the acting player's move will work, fail, or have other unintended consequences.[4] Next, like the "Dungeon Master" in the classic entertainment game *Dungeons & Dragons*, a facilitator weighs the presented arguments and assigns a probability for the action to succeed as intended (e.g., a dice roll of eight or higher using two six-sided dice, which is equivalent to a 42 percent chance of success). The acting player then rolls dice to determine if their proposed action is successful or not.

Next, the facilitator describes the effects of a player's action on the game world through creative—but ideally realistic—storytelling. Through these arguments and stories, the players and facilitator build a shared narrative about the game world and the factors that influence its development over the course of play. Structured arguments and probabilistic adjudication advance the game state or sequence, allowing players to both experience and discuss how their actions affect the narrative over time.

The original version of *Pinnacle Protagonist* in 2017 used this standard matrix game format with two minor modifications. First, we created a map, game pieces, and basic trackers (e.g., "the world economy") to help players remember what had occurred thus far. Second, we included a mechanism for players to take secret actions. Players who wished to make secret moves (i.e., actions that would not be immediately revealed in open discussion) would step aside with the facilitator on their turn, describe their action, argue for it, and then roll the dice without the usual time for counterarguments by other players.

DESIGN CHALLENGES

For our game, groups of three or four students were assigned to one of several country teams. These country teams were the United States, Russia, Western Europe (i.e., the United Kingdom, Germany, and France), the Nordic states, and a combined team for the Baltic states and Poland. The scenario posited Russian election interference and cyberattacks against Latvia, leading to escalatory actions by the various powers in the region.

The original version of *Pinnacle Protagonist* was well received by students and faculty. Nevertheless, after two years of experience running this matrix game and similar tabletop exercises, we realized that the original format was not adequately capturing two key aspects of strategic cyber and information warfare: (1) anonymity and attribution, and (2) the resource management of cyber tools.

Anonymity and Attribution

The basic format of matrix games is not well suited to exploring some aspects of cyber detection, attribution, and deception. In our games, players attempted to take "nonattributable" actions against other players at the open table as regular moves and by using the secret move mechanic. Unfortunately, it is difficult for players to suspend belief enough to ignore which team has left the room and how subsequent events negatively affected other teams. Most players realized that an allegedly unattributable action must have come from the team that left the room with the facilitator, and they acted accordingly. Since most developments in the game are generated by the players and facilitator, there is essentially no noise for secret moves to hide in.

Despite this challenge, we still wanted to include the interplay between anonymity and attribution in our game mechanics. We hoped to faithfully replicate this interactive feature of cyber warfare for both offense and defense. As Aaron Brantly argues, "The ability for states to respond to cyberattacks and their perpetrators is an important aspect of cyber defense. But when we consider the offensive strategy of a state, the concern is a combination of achieving the objective and avoiding attribution." In other words, "anonymity is a fundamental aspect of offensive cyber operations," and "attribution is a fundamental aspect of cyber defense."[5]

Cyber Resource Management

Without modification, players' actions in the standard matrix game format poorly approximate the long lead times and single-use tools in the cyber domain. Players may face soft limits in the form of counterarguments and the number of actions they can attempt over the course of a matrix game, but arguments about resource availability are usually clumsy. The caution that drives real decision-makers to shepherd long-lead-time, single-use tools is hard to capture in a game where players may only experience a few rounds of play.

In the real world, nations have limited resources. Therefore, we wanted to force the players to think through their strategies and create a small number of tools with strategic effects. In doing so, student players would have to anticipate their future needs and understand or appreciate the trade-offs and risks. Moreover, as described in chapter 6, developing effective cyber and information tools often requires long-term investments of time and resources. Success requires strategic thinking, planning, and resource prioritization.

A famous real-world example of this long-lead-time tool is the Stuxnet virus, which was used to attack Iranian industrial sites, including a uranium-enrichment plant at Natanz.[6] According to the computer security firm Kaspersky, "a team of 10 people would have needed at least two or three years to create" Stuxnet.[7] According to a similar assessment by the computer security firm Symantec,

> [The] code is sophisticated, incredibly large, required numerous experts in different fields, and mostly bug-free, which is rare for your average piece of malware. Stuxnet is clearly not average. We estimate the core team was five to ten people and they developed Stuxnet over six months. The development was in all likelihood highly organized and thus this estimate doesn't include the quality assurance and management resources needed to organize the development as well as a probable host of other resources required, such as people to set up test systems to mirror the target environment and maintain the command and control server.[8]

The allocation of time and personnel resources to develop Stuxnet implies strategic decision-making and resource management choices.

THE REVISED DESIGN

To address these challenges, we surveyed other cyber wargames and exercises that dealt with strategic cyber and information warfare. We found several professional cyber wargames used in government and industry, but many focused on the technical and tactical level. Of the few games and exercises we encountered that examined cyber and information warfare at the strategic level, fewer still had mechanics suitable for use in a matrix game within our time and classification constraints. Of course, it is very possible that there were relevant games that we did not find at the time; project deadlines and limited access to wargaming in the defense and intelligence communities prevented an exhaustive search.

Fortunately, we found several recreational games with useful mechanics. Social deduction games, which are sometimes referred to as hidden identity games (e.g., *Mafia* and *Werewolf*), feature innovative ways for players to practice deception while avoiding detection and attribution in an open table setting. We also drew inspiration from the classic board game *Diplomacy*, which focuses on players' manipulation of the information environment.[9] Finally, we examined games that require players to develop and allocate resources with limited knowledge of future circumstances. We were particularly drawn to games with deckbuilding or engine-building mechanics, such as *Race for the Galaxy* and *Magic: The Gathering*, whereby players assemble a toolbox of capabilities before and during gameplay to outcompete their opponents.[10]

Drawing on this inspiration, we revised our basic design. These changes involved a new adjudication tool and alterations to the matrix structure. We developed a semirigid tool to help adjudicate strategic cyber and information operations that is easy to use and sufficiently generic for an unclassified wargame. Although semirigid adjudication meant forgoing the benefits of structured argument in some parts of gameplay, the fidelity gained with this tool was worth it. To craft this tool, we needed a typology of targets and effects for cyber and information operations, but no satisfactory and unclassified typology existed.

Our game tool became the National Strategic Program (NSP) framework, as shown in figure 14.1. Creating this new adjudicative tool also allowed us to alter the round-based structure of the matrix game to better address covert actions. Surveying secret actions in other games, we found one of the most common and robust methods was to have all players act simultaneously—simultaneous action provided for plausible deniability.

The Road to the NSP

We methodically developed our NSP framework. We began by describing a list of high-level targets that players might want to influence or attack in a strategic wargame. Given the wide range of potential targets, we needed to put them in manageable categories, as illustrated in figure 14.1, section 1 (i.e., opinion, civilian critical infrastructure, military systems, governance, and business). For

NATIONAL STRATEGIC PROGRAM DEVELOPMENT FRAMEWORK

EXAMPLE

DESCRIPTION ⑥

TARGET COUNTRY ④ — **MODIFIER TO BASELINE FOR SUCCESS**

TARGET COUNTRY	MODIFIER TO BASELINE FOR SUCCESS
United States Russia	0
Western Europe	0
Baltic States and Poland	-1
Nordic States	+1
	-1

CATEGORY	TARGET SECTOR	BASELINE MINIMUM ROLL FOR SUCCESS / AVOID ATTRIBUTION
Opinion ①	General public opinion	6/5 ②
	Government decision makers' opinion	6/5
	Elite Non-government public opinion	5/5
Civilian Critical Infrastructure	Civilian critical infrastructure - information / communication	7/8
	Civilian critical infrastructure - energy	7/8
	Civilian critical infrastructure - financial	10/7
	Civilian critical infrastructure - transportation	9/7
	Civilian critical infrastructure - emergency services	10/7
Military Systems	Military - ISR	6/9
	Military - Information / communication	8/7
	Military - weapon systems (non-nuclear)	9/8
	Military - Logistics / transportation	9/8
	Military - Other	6/9
Governance	Governance - election systems	7/8
	Governance - operations	8/8
Business	Defense industrial base	8/9
	Civilian commercial entities	8/9

DESIRED EFFECT ③ — **MODIFIER TO BASELINE**

DESIRED EFFECT	MODIFIER TO BASELINE
Physical Destruction of System	+ 4 Success Min Roll requirement
Disrupt / Deny Service or Capability	0
Persuade/ influence Opinions	0
Exfiltrate Data	-1/-1 vs Business Category Sectors
Corrupt Data	-1/ -1 vs Civilian Critical Infrastructure Sectors

FACILITATOR USE ONLY

MIN ROLL FOR SUCCESS   ⑦   MIN ROLL TO AVOID ATTRIBUTION

PROGRAM SUPPORTED BY (- 2 SUCCESS/+1 ATTRIBUTION): ⑤

______US ______RS ______WEU ______BSP ______NS

FIGURE 14.1. Sample of the National Strategic Program Development Game Sheet

a baseline of the NSP framework, we chose the emergency support functions, as outlined in the US National Response Framework.[11] These functions are an organizing principle that the Department of Homeland Security uses to share information with the private sector and coordinate the federal response in a national emergency. They were a logical choice for our game.

The US government has designated sixteen critical infrastructure sectors "whose assets, systems, and networks, whether physical or virtual, are considered so vital to the United States that their incapacitation or destruction would have a debilitating effect on security, national economic security, national public health or safety, or any combination thereof."[12] The last few categories of targets include different military capabilities, public opinion, and governance systems. Our addition of these categories, plus our consolidation of emergency support functions and critical infrastructure sectors, led to five broad categories and seventeen more specific target sectors. These targets are not based on the specific technologies involved (e.g., supervisory control and data acquisition systems). Rather, they are based on the importance of these sectors to the functioning of a modern society and military. In our view, a strategic game does not necessarily require a greater level of technical detail.

The next question was how to represent the probabilities of successfully attacking a target sector and avoiding attribution (see figure 14.1, section 2). This is where our design deviates from a traditional matrix game. In the traditional matrix game format, the probabilities of success are created by the give and take of structured arguments, either with other players for public actions or with the facilitator for secret actions. In contrast, using the NSP framework, secret actions in the cyber and information domain are adjudicated using a baseline set of predetermined probabilities. Then, the baseline probabilities can be modified by subject matter expertise and player arguments to the facilitator or umpire. This option retained players' incentives to develop reasoned courses of action.

The baseline probabilities of success and avoiding attribution are also different for each country team. These differences represent each country's comparative advantages regarding access to technical skills. They also take into account historic practices (e.g., Russian history and experience in using misinformation), legal or institutional constraints (e.g., high legal and ethical barriers to acting against certain target sets), and cultural affinity (e.g., general favorable global opinion toward many of the Nordic states). It was not possible to perfectly represent each country's real-world capabilities (especially not in an unclassified game). Therefore, probabilities were assigned based on a broad estimate of relative national strengths and weaknesses, which we derived from publicly available information, in consultation with our sponsor.

The odds of avoiding attribution followed a similar logic. Teams also have the option of deliberately allowing their actions to be attributed to them. If a team's strategy involves executing their NSP and letting the target country to know it was them (e.g., for strategic signaling), then the attribution roll is not conducted.

Once players have identified a target and the probabilities of success, they need to achieve something from their attack. We looked at real-world cyber incidents and attacks to identify realistic effects. The effects were boiled down to five categories, illustrated in figure 14.1, section 3: (1) physical destruction of systems; (2) disrupt/deny service or capability; (3) persuade/influence opinions; (4) exfiltrate data, and (5) corrupt data. By focusing on the effects of the attacks instead of the underlying technical causes, we minimized the risk of classification issues. Here again, we created country-team-specific modifiers that reflected comparative advantages in achieving the desired effect.

Once the issues of target sectors and effects were addressed, we added the other player teams (figure 14.1, section 4) as part of the target designation process. Modifiers were added to reflect the nature of the relationship and history between the countries represented by player teams. Deep historic connections, integrated economies, and military alliances can improve the chances of success. Conversely, historic animosity, a lack of business and cultural ties, and military rivalries can decrease the chance of success.

At the request of our faculty sponsor, we added a mechanism that allowed players to work with other player teams when preparing their NSP (figure 14.1, section 5). This mechanism reflected gains from the coordination and collaboration with partners and allies, while also acknowledging the opportunity costs of cooperation and increased chance of attribution for combined or coalition action. For game simplification, teams could only have one partner on any NSP.

The NSP checklist described so far provides a basic outline of players' actions. However, to develop a rich and convincing narrative, the facilitator needs more details from the players about their actions and intentions (figure 14.1, section 6). Then, after player teams have made their choices, they hand their NSP over to the facilitator. The facilitator adds up the chances of successful execution and avoidance of attribution (figure 14.1, section 7). The facilitator's review finalizes the NSP sheet.

Integrating the NSP into Gameplay

The next key decision was to figure out how to integrate the NSP tool into the overall game flow. *Pinnacle Protagonist* was a three-day-long event. On day 1, players were introduced to the matrix game system and participated in a practice session. The player teams then developed their respective NSPs on day 2. To reflect resource constraints, most country teams were allowed to create three NSPs, but they then had to choose and execute only one during gameplay. (To balance game dynamics and reflect assumed comparative advantages with cyber and information warfare, the Russian team was allowed to create four NSPs and execute two during gameplay.) The timing of NSP creation forced students to develop strategies at the outset while thinking through associated branches and sequels in preparation for day 3 of gameplay. As intended, the limited number of NSPs forced students to think carefully about how they wanted to use finite resources.

Figure 14.2 depicts how NSPs were incorporated into gameplay on day 3. At the beginning of this round, there was a ten- to fifteen-minute negotiation phase. Player teams discussed and coordinated their actions for the open action phase. After this initial discussion, each team was brought to a breakout room to meet with the facilitator. During this meeting, the team declared whether they would execute one of their NSPs. If the team did not want to execute an NSP, they waited a minute or two to avoid arousing suspicion before returning to the plenary room. This process was repeated for each team.

If the team members wanted to execute an NSP, they presented their NSP sheet to the facilitator. The team and facilitator quickly reviewed the NSP to ensure that the facilitator was clear about the team's intentions. The team then rolled dice to determine if their NSP succeeded and achieved the desired effects.

GAME FLOW

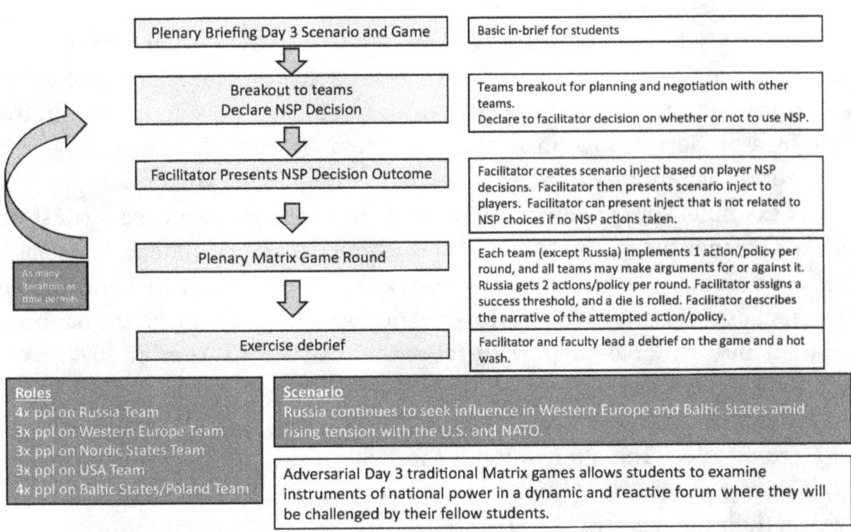

FIGURE 14.2. Game Flow (Russia Example)

Then, after the team left the room, the facilitator rolled the dice to determine if the team in question avoided attribution. The facilitator also took notes on what effects the team's actions had on advancing the game's narrative. The next team was then called to the facilitator and went through the same process without knowing what had happened to the previous team.

After all the teams played or passed on executing their NSP, the facilitator crafted a narrative and scenario inject or update based on the results. If no team executed an NSP, the facilitator created an inject to reflect the regular random events that might occur in the real world. The facilitator's inject kept the players guessing as to whether events were deliberately generated by another team. As a result, the anonymity of players' actions was maintained through this minor change in the game process. After the facilitator announced the result of the NSP decisions, the teams conducted open or "public" actions using the standard matrix format of structured argumentation and collaborative or community adjudication.

LESSONS LEARNED

Overall, our NSP tool added another layer of depth and sophistication to gameplay beyond what the standard matrix format provides. We found that the process of players developing and using their NSP created correspondingly new dynamics when compared with the previous iterations of *Pinnacle Protagonist* without the NSP.

We made two additional observations. First, process is as important as any other tool used in a wargame, cyber or otherwise. Critical features of cyber operations—anonymity and the attribution problem—were replicated in the game by a modest change in wargame process. This small change created a degree of realism not seen in the standard matrix game format.

For example, some teams created NSPs to potentially be used against an allied or partner team. The Nordic states team created an influence campaign to be used on the Western European team as a hedge—a move that anticipated important aspects of the real-world influence operations that Ukraine has used to garner international support to resist the Russian invasion in 2022.[13] We would not expect this type of action in open play because such a capability would likely be viewed with suspicion and distrust by other teams aligned with the Nordic states. The anonymity of the NSP process enabled the members of the Nordic states team to consider and develop a hedge where they thought it was in their interests. In this way, students demonstrated a far more nuanced application of their team's national interests and strategy.

Second, educational cyber wargames can address some of the ways that resource management shapes players' strategies by leveraging wargaming's focus on human decision-making under uncertainty. In a standard matrix game, players are limited by the finite number of actions they can take over the course of a session. This gameplay dynamic forces players to act strategically, but it can also incentivize players to stretch the plausibility of their actions, since they can attempt to do anything they are willing to argue is plausible. This is a potential problem for strategic cyber games, where it is relatively easy to argue that a country "could have" developed the capability it needs in any given turn (and that it could have had the strategic foresight to do so).

Our design incentivized players to think more carefully about using their cyber capabilities. The NSP requires players to craft specific courses of action ahead of time and limits their use during a game. For example, exploiting cyber access to an adversary's critical infrastructure now may mean burning the possibility of reusing that access again in the future. But delay was not always advantageous. On more than one occasion, player teams held off on executing an NSP in order to time their single-use actions for maximum effect, only to discover that another team's actions had already made their cyber access or exploit irrelevant. This happens in the real world. Thus, while the constraints of the NSP process were artificial, they nevertheless produced realistic pressures and constraints on players' decision-making during the game.

Granted, our implementation of the NSP tool was not without its flaws. One potential shortcoming is that we did not explicitly force players to act upon NSP outcomes. Rather, we left it up to the player teams to argue how the effects of their NSP might have an impact on other actions; for better or worse, this left the players with wiggle room to argue their way around subsequent effects and unintended consequences. Another potential drawback of our design is that, like the secret move mechanism it replaced, the NSP is adjudicated outside the normal process of open argumentation between the players, which is the heart of matrix gaming.

In the end, the NSP tool and process enhanced the educational gaming experience. Feedback from student evaluations and faculty commentary were overall very positive. For example, one student wrote, "excellent metrics and ruleset development for nonlethal effects generation; . . . [the] best I have seen in wargaming!" We retained most of the benefits of structured argumentation in matrix gaming while adding a more nuanced and realistic representation of strategic cyber and information warfare. Developing their NSP required player teams to think and plan in anticipation of future requirements and the finite resources available. Executing their NSP created uncertainty and anonymity, providing players with freedom of action amid a sea of doubt. All this was accomplished without requiring an increase in control staff and only a nominal increase in time required to adjudicate. We plan to continue using this format, and we will keep working to better integrate NSP-generated events into gameplay in future iterations of *Pinnacle Protagonist*.

NOTES

1. Among others, see US Air Force, *Challenges of Cyberspace Operations*, Air Force Doctrine Publication 3-12 (Arlington, VA: US Department of Defense, 2011), www.doctrine.af.mil/Portals/61/documents/AFDP_3-12/3-12-D04-CYBER-Challenges.pdf.

2. US Department of Defense, "Officer Professional Military Education Policy," Chairman of the Joint Chiefs of Staff Instruction 1800.01F, May 15, 2020, www.jcs.mil/Portals/36/Documents/Doctrine/education/cjcsi_1800_01f.pdf?ver=2020-05-15-102430-580.

3. Chris Engle, "The Classical Matrix Game," 2002, https://drive.google.com/file/d/1VB2QuWh53qVE_KWsQs9fvNMpMsGj89HM/view.

4. Adjudication through argumentation in matrix games not only relates to the age-old Socratic method of teaching but also contemporary debates about the role of debate in international relations. E.g., see Thomas Risse, "'Let's Argue!': Communicative Action in World Politics," *International Organization* 54, no. 1 (Winter 2000): 1–39.

5. Aaron Franklin Brantly, *The Decision to Attack: Military and Intelligence Cyber Decision-Making* (Athens: University of Georgia Press, 2016), 87.

6. The extensive reporting on Stuxnet includes John Markoff, "Malware Aimed at Iran Hit Five Sites, Report Says," *New York Times*, February 11, 2011.

7. David Kushner, "The Real Story of Stuxnet," *IEEE Spectrum*, February 26, 2013, https://spectrum.ieee.org/telecom/security/the-real-story-of-stuxnet.

8. "Migration User," W32.Stuxnet Dossier blog post, October 1, 2010, https://community.broadcom.com/symantecenterprise/communities/community-home/librarydocuments/viewdocument?DocumentKey=1a29f93b-41f8-4265-a7d1-44c3c94d3b52&CommunityKey=1ecf5f55-9545-44d6-b0f4-4e4a7f5f5e68&tab=librarydocuments.

9. Allan B. Calhamer, *Diplomacy*, 4th ed. (Baltimore: Avalon Hill, 2000), board game.

10. Thomas Lehmann, *Race for the Galaxy* (Rio Ranch, NM: Rio Grande Games, 2007), card game; Richard Garfield, *Magic: The Gathering* (Seattle: Wizards of the Coast, 1993), card game.

11. US Department of Homeland Security, *National Response Framework*, 4th ed. (Washington, DC: US Government Publishing Office, 2019), 39–41, www.fema.gov/media-library-data/1572366339630-0e9278a0ede9ee129025182b4d0f818e/National_Response_Framework_4th_20191028.pdf.

12. "Critical Infrastructure Sectors," Cybersecurity and Infrastructure Security Agency, www.cisa.gov/critical-infrastructure-sectors.

13. Daniel Grobarcik, "Snakes, Stamps, and Javelins: How Ukrainian Information and Influence Operations Brought the Fight to Russia," *Irregular Warfare Initiative*, June 15, 2023, https://irregularwarfare.org/articles/snakes-stamps-and-javelins-how-ukrainian-information-and-influence-operations-brought-the-fight-to-russia.

PART III

CONCLUSION

15

BALANCING ZEALOTS AND SKEPTICS ABOUT EMERGING TECHNOLOGY: A WARGAMING BLUEPRINT FOR INNOVATION

NINA A. KOLLARS AND BENJAMIN H. SCHECHTER

What does cyber wargaming reveal about how people think about innovation and emerging technology? The preceding chapters have attempted to pull back the veil on cyber wargaming, presenting an array of games that students, scholars, and practitioners can use to understand cyberspace. As the earlier chapters have demonstrated, cyber wargaming is maturing rapidly as a tool in the classroom, the boardroom, and the war room. Our overarching objective has been to make cyber wargaming less mysterious and more accessible as a tool for making sense of cyberspace and its security implications.

This final chapter explores the insights that cyber wargaming offers for understanding emerging technologies and the associated potential for innovation. We posit that the lessons learned from cyber wargaming hold promise for a whole host of technological trends, as what we consider "cyberspace" is an amalgamation of once-emerging and, in some cases, still-emerging networks of digital technologies. This book illustrates how wargames grapple with understanding the digital domain. We contend that many of these lessons apply more broadly, providing valuable insights for wargaming other complex and challenging questions about new technologies.

In this chapter, we seek to answer the question: What is the value of wargames when it comes to emerging technologies and innovation? We organize our discussion into four sections. First, we argue that the primary value of wargaming lies in its ability to help us make sense of poorly understood technologies by orienting us toward potential applications, actions, and effects. Wargaming, for this purpose, is as much a journey of discovery as it is a tool for more rigorous, scientific research and analysis. Second, we maintain that effective wargames strike a careful balance between two positions on the likelihood of technological innovation: zealotry and skepticism. This balance is crucial for enabling sensemaking about technological systems that are not wholly realized.

Third, we emphasize that well-balanced wargames, which embrace a logic of discovery, are not merely exercises in studying inevitable or deterministic "truths" about future technology. They also serve as potential pathways to shape if not create technologies and associated practices for machines that do not yet exist. As such, some wargames about emerging technology are potentially productive or generative of the very things they purport to explore, making them potential enablers of interlocking social and technical change. Fourth and finally, we stress the importance of recognizing this potential, which should not be taken lightly.

WARGAMING TECHNOLOGICAL FANTASIES AND ANXIETIES

The concept of emerging technology—particularly as it pertains to national power—is in vogue in contemporary discourse and practice about security. The pervasive effects of digitization and computation on our daily lives seems to have only intensified interest in nascent technological developments. Quantum computing, artificial intelligence (AI), and autonomous systems are now common examples of emerging technologies with potentially profound implication for national and international security. Notably, these technologies also depend on cyberspace, directly or indirectly, and thus share the complex and dynamic nature of this digital domain.

The implications of emerging technologies are inherently unclear (i.e., uncertain and indeterminate). The ambiguous nature of these technologies provokes both fascination and apprehension about potential change and disruption.[1] These feelings can likewise leave us with the sense that, in the words of Langdon Winner, "technics" are out of control.[2] There lurks a tacit determinism in much of the discourse surrounding technological change that, unfortunately, fails to situate new technologies within the context of existing social, political, and economic structures, as well as agency and choice. Consequently, people sometimes feel vulnerable and powerless in the face of ill-defined and supposedly inexorable change, making them potential casualties of their own inventions. But they also adapt.

Recent responses to one generative text AI, ChatGPT, illustrate the range of reactions to emerging technologies.[3] Among other potential applications for education, this large language model can generate text on demand to answer high school and even college level term paper questions, using language that seems natural if not accurate. The initial response among some educators was panic, with faculty and administrators speculating that cheating by students would become even more rampant than it already is, and the ability to accurately assess educational outcomes would be lost.

However, some unrest subsided and gave way to naysayers who, upon closer examination, argued that it is still possible to distinguish between human-generated and AI-generated text (for now). Opportunistic behavior also emerged, such as a one young man who created an online course and quickly earned over $35,000 teaching people how to use the generative tool.[4] Faculty are also learning how to leverage ChatGPT to enhance student education in the classroom.

Similar reactions to emerging technologies—from ChatGPT to hypersonic missiles—are evident in national security communities. Here anxiety arises not just from the fear of losing control but also from apprehension over who might gain it. Power is distributed in international relations, and there is a tendency to equate emerging technologies in particular with military power. The belief that new technologies will inevitably change the nature of warfare is widespread. By extension, one common assumption is that being the first to develop or "harness" emerging technologies is essential for national power and security.[5]

From this perspective, technological change can appear to condone an aggressive stance in a relentless technological arms race.[6] Since the distribution of power is relative, national security communities often focus on maintaining or gaining more control over technological change than potential adversaries. Thus, the logic and language of technological competition tend to revolve around controlling and exploiting emerging technologies to avoid becoming a victim of them.

Wargaming can help us make better sense of these dynamics by critically examining the drivers of our anxieties, exploring potential opportunities, and defining the art of the possible for conflict as well as cooperation. Wargames help us study the potential implications of technologies that are purported to be transformative but are not yet well understood. In this sense, wargaming is about sensemaking, which includes making sense of things that do not exist and the potential conditions for creating and using them.

MAKING SENSE OF IMAGINARY TECHNOLOGY

The concepts of sensemaking, emerging technology, and innovation play critical roles in understanding wargaming and its relationship to the development and use of new knowledge and tools.

For its part, "innovation" is a term that is often overused, misused, and outright abused, from Silicon Valley to the Beltway. We use it to refer to successful, intentional, and systemic change. For instance, when a factory transitions from individual craftsmen to workers on a production line, this change exemplifies innovation. Technological innovation, then, is the intentional and successful change brought about by the adoption or development of new technology. This is an intentional process—slow or rapid, diffuse or centralized—of designing and integrating new tools and knowledge into new and extant institutions. Unlike the rhetorical use of "innovation," which can imply both improvement and "disruption," our definition does not place normative judgement on the effectiveness or enhancement of a system, but simply notes that the change was deliberate and successful.[7]

When discussing "emerging technology," we specifically refer to technologies that are *perceived* or expected to have the potential to transform sociotechnical systems.[8] The terms "perceived," "expected," and "potential" carry significant weight. Emerging technologies are considered as such because their ultimate purposes and implications remain uncertain. The discourse surrounding emerging technologies often features elements of technology hype, marked by exceptional expectations fueled by the indeterminate and unpredictable nature of future technology.[9]

For example, electricity was initially seen as an emerging technology with uncertain applications. It was greeted with skepticism and anxiety. Through invention and zealous advocacy, electricity transitioned from a novelty to a technology with widespread adoption and adaptation. With the benefit of hindsight, we can now credibly categorize it as transformational. But this outcome was not inevitable. History could have been different.

The term "sensemaking" refers to the process of orienting individuals within a system so that their actions and intentions are understood by others in that system.[10] As poetically described by Weick and Sutcliffe, sensemaking "begins with chaos, and resolves in action."[11] It is the process through which individuals become oriented to an environment and learn how to navigate it. Sensemaking happens anytime humans confront new systems. The concept can be applied to a wide variety of situations, such as traveling to a foreign country and trying to understand the local customs and rules.

In wargames, sensemaking follows a similar pattern. First, players familiarize themselves with the game. This involves reading instructions, talking with other players, and understanding the game's rules and mechanics. Next, players become oriented to the game's abstracted world, gradually gaining a deeper understanding of the fictional scenario, its players, and component parts. Ultimately, the wargame allows players to apply their understanding of the game world to real world scenarios. Sensemaking in wargames is a complex process that often goes unrecognized.[12] However, the mechanisms by which sensemaking occurs through signs, signals, explanations, experiences, and interactions are integral to the wargaming experience.[13] The mechanisms devised to facilitate sensemaking in cyber wargames, and discussed throughout this book, also have useful applications to sensemaking in games about other emerging technologies.

ZEALOTS, SKEPTICS, AND THE CURIOUS OBSERVER

Effective wargaming vis-à-vis emerging technologies entails a gradual, scenario by scenario, exploration of fictional worlds that could not only herald transformation and innovation but also marginal change, if not stasis. We believe that successful games strike a careful balance between dueling perspectives on technological innovation. On one hand is zealotry, which is future oriented and focused on possible outcomes. On the other hand is skepticism, which is typically past or present oriented and focused on the most likely outcomes.

Zealots are characterized by their strong belief in the potential for transformational technological change. More than just technophiles, zealots can creatively contribute to peoples' ability to envision sociotechnical landscapes that do not exist (at least not yet). They are biased toward change; their preferences for outcomes that are not yet actualized but could be fuels their beliefs.

Zealots may extol the potential value of an emerging technology just as vigorously as they preach its potentially negative or outright harmful consequences. For example, early zealots for cyberspace voiced utopian ideals of a more decentralized, open, and free space to share information.[14] At the same time, zealots for cyber war argued that cyber capabilities could enhance conventional kinetic fighting or lead down the path to nuclear war through unchecked digital escalation.[15] As in Silicon Valley, zealots in the defense industry often promise the revolutionary potential of their products, possibly motivated more by profit than ideology. The rise of fiction intelligence—also called "useful fiction," or science fiction intended to help conceptualize future trends—can similarly be seen as part of the zealot camp, as it helps military personnel and civil servants envision future scenarios absent the inconveniences of reality.[16] In wargames, zealots can help overcome status quo biases, explore possibilities unrestricted by current limitations, and promote a vision of "what could be."

In contrast to zealots are skeptics. Skeptics often consider themselves realists. They are oriented toward the past, present, and most likely outcomes. This typically means that skeptics are more contextual, recognizing the limitations presented by entrenched technologies and practices, biased toward observable data rather than promise and potential. Skeptics are not necessarily Luddites or defenders of the status quo (although some certainly are). But they acknowledge the inertia that innovation must overcome.

Skeptics may be driven by a commitment to intellectual rigor, cynicism, or outright self-interest. Many academics are often tacit skeptics because of their commitment to hypothesis testing, data, and falsification.[17] Within the Department of Defense, the same may be true of military and operations research analysts, who emphasize data-driven decision-making and often take a skeptical stance. These skeptics are leery of the uncertainty and indeterminacy of emerging technology. Veteran practitioners also tend to be skeptics. Many are all too familiar with the resistance to change in large institutions and the gap between engineers' intentions and real-world implementation. Military operators, in particular, may be especially wary of relying on untested technologies for their careers or lives. History has been a harsh teacher of the risks of relying on unproven gear. Skeptics may scoff at the perceived naïveté of some zealots and defense contractors' slick marketing brochures, but some may also be seeking to promote or sell traditional, established, or entrenched solutions and equipment. Skeptics, like zealots, are biased. But their biases tend toward the limits of change and lower likelihoods of technological innovation.

Successfully balancing the biases of zealots and skeptics in wargaming scenarios can help participants experience the process of sensemaking in a more comprehensive and nuanced way. By incorporating both perspectives, wargames can generate more diverse and robust outcomes, enhancing their ability to explore the implications of emerging technologies. The best technology wargames, like cyber wargames, need elements of both zealotry and skepticism to create scenarios and gameplay that incorporate the excitement of technological potential along with the constraints of reality.

Games with Zealots and Skeptics

A bias toward the zealot's perspective on wargames allows for broad exploration and unfettered creativity, engaging players in the discovery process without the constraints of the status quo. You should be prepared for *Star Trek* options—exceedingly creative but unrealistic capabilities—wherein players may imagine fantasy capabilities that are not likely to exist anytime soon, if ever.[18]

To create a wargame oriented toward the zealot, simplicity of game design and relaxation of technical rules are ideal. Designers should reduce the level of details players need to understand specific technologies. This simplification can reduce players' cognitive load and allow them to focus on the larger operational or strategic picture. In the case of cyberspace, some cyber capabilities evolve faster than conventional capabilities, meaning that they can quickly go from cutting edge to outdated or irrelevant. Moreover, making the wargame extremely technical can make players think that they need to make decisions of comparable specificity, which may further ground them in the here-and-now of technology. Leaning toward zealotry may be controversial, since it can produce unfounded expectations. Nevertheless, it can also reveal creative possibilities that lead to new and even surprising understandings of how emerging technologies can be used.

For skeptics, the value of wargaming typically focuses on answering concrete questions rather than indulging in flights of fancy. Wargames that favor skepticism emphasize sensemaking for students, leaders, and analysts. They focus on learning from the wargame by imparting specific and structured knowledge. In an educational context, the skeptic's position is advantageous for some learning outcomes. Wargaming in the classroom can enhance understanding of the complex nature of cyberspace and emerging technologies, provided the approach is clear and rigorous. Skeptical games ground exploration in the richness of the real-world context, emphasizing the often slow and uncertain nature of change through technology, which is neither immediate nor assured.

It is essential to balance zealotry with skepticism to maintain the integrity and educational or analytical value of the wargame. Games can lose their value and become unnecessarily tedious or merely entertainment without a proper balance. Currently, there is an unhealthy separation between pedagogical scholarship and professional wargaming. Bridging this gap is essential to effectively leverage wargames as a cornerstone of active learning research.[19]

Achieving a balance between zealotry and skepticism in wargame design is particularly important for effectively exploring the implications of emerging technologies.[20] Unchecked gameplay can result in ethical dilemmas, misunderstandings of technologies and systemic change, and undue influence on players' perceptions. For example, the potential influence of profit-seeking is high in some cyber wargames, since firms that sell cybersecurity tools may also participate in design and play.[21] Hosting such a game can be a mechanism for marketing hardware, software, and service solutions. Because wargames provide players and observers with what feels like an exploratory look into future worlds, how the game environment and tools are represented can have an overly determinative effect on players' sensemaking. Zealots or skeptics pushing an agenda through a wargame can affect play, outcomes, and lessons learned, which can affect how players and policymakers think about the real world.

By balancing zealotry and skepticism in wargame design, players can meaningfully explore emerging technologies while maintaining a strong connection to real-world constraints. This balance ensures that wargames serve as effective tools for research and education, fostering a deeper understanding of the complex interactions between technology and society, including warfare. Encouraging a thoughtful approach to balance these perspectives will not only create a more engaging and realistic gaming experience but also result in valuable insights and practical applications for the development and deployment of emerging technologies.

MAKING WHILE PLAYING

Wargames that explore emerging technologies have the potential to influence not only our understanding of these technologies but also their subsequent development and application. Here again, the balance between zealots and skeptics is important: by shaping our expectations about potential technological futures, wargames can also shape what we do as a result.

Just as players make decisions about conflict, cooperation, and the consequences of their actions in a wargame, they also make important decisions in real life. This is most apparent when the players are c-suite executives or government leaders and policy elites. But it is also true for students who may fill these roles in the future.

As wargames help these players make sense of potential futures, their preferences and beliefs can affect decisions about those futures. Such decisions range from how and where real research dollars are spent to what policies and laws are pursued or eschewed. In this sense, wargames about emerging technology are not merely tests about some objective or inevitable truth; they are also opportunities to shape technological futures as a consequence of exploration through the game. They help shape consequential perceptions and expectations that can in turn shape the future.

The generative, constructive, and consequential aspects of wargaming emerging technologies for real life can mean that they are not a purely neutral or apolitical exploration. Wargames that involve policymakers, military leaders, and business executives can influence what powerful people think they can, should, and will do with emerging technology. Though perhaps more indirect, they can also have a similar influence on other audiences, from scientists and engineers to students.

Innovation does not occur in a vacuum, and science and technology do not emerge or evolve independently of human context and imagination. Our expectations about the future potential of supposedly transformational technologies can affect technology itself, as well as the social, political, and economic processes of translating that technology from ideation to emerging to actualized. These potential consequences can be advantageous but also problematic—especially if done without careful consideration. What we do and learn from these games can influence what we do with technology in the real world, even before that technology is fully realized in the real world.[22] Or, said another way, wargaming emerging technologies can affect technological innovation.

MORE COLLABORATION AND PRACTICE ARE GOOD

Wargames are valuable for both zealots and skeptics. The case of cyber wargaming serves as a model for how wargames can inform discovery and influence the development of emerging technologies—for good or ill. As players, students, and analysts engage with these games, their insights can shape and inspire real decisions. Having more diverse communities of thought involved and invested in wargaming stands to help us make better decisions as a result.

Of course, wargaming emerging technologies is not without challenges and pitfalls. Balancing zealotry and skepticism, maintaining ethical boundaries, and ensuring meaningful exploration of emerging technologies takes work. Working to overcome these challenges requires us all to cultivate an environment of critical reflection, maintain a clear focus on educational and research objectives, and encourage open dialogue among participants. It also requires wargaming and technology communities that are dedicated to critical self-reflection, information sharing, and improvement—which this book hopes to help foster. In the rapidly evolving landscape of digital and other emerging technologies, continuous learning will be essential.

The value of learning through wargaming lies at the heart of this volume. We hope it serves as a model and inspiration for further collaboration. In addition, we make a final request for more practice with wargaming about cyber and other emerging technologies. Reading a book about riding bicycles is one thing, but riding a bike yourself is a very different experience. More practice playing wargames is good, regardless of your experience and expertise. This is true now, and we suspect the same for wargaming in the future.

NOTES

1. Joel Mokyr, Chris Vickers, and Nicolas L. Ziebarth, "The History of Technological Anxiety and the Future of Economic Growth: Is This Time Different?" *Journal of Economic Perspectives* 29, no. 3 (2015): 31–50.
2. Langdon Winner, *Autonomous Technology: Technics-Out-of-Control as a Theme in Political Thought* (Cambridge, MA: MIT Press, 1978).
3. Kevin Roose, "Don't Ban ChatGPT in Schools; Teach with It," *New York Times*, January 12, 2023, www.nytimes.com/2023/01/12/technology/chatgpt-schools-teachers.html.
4. Aaron Mok, "A 23-Year-Old with No Technical Training Earned $35,000 in 3 Months Teaching ChatGPT to Newbies," *Business Insider*, April 27, 2023, www.businessinsider.com/how-to-earn-money-teaching-chatgpt-course-to-beginners-udemy-2023-3.
5. Office of the Director of National Intelligence, "The Future of the Battlefield," in *Global Trends: Deeper Looks* (Washington, DC: Office of the Director of National Intelligence, 2021), www.dni.gov/index.php/gt2040-home.
6. Ajey Lele, "Quantum (Arms) Race," in *Quantum Technologies and Military Strategy* (New York: Springer, 2021), 145–72.
7. In adopting this definition, we are borrowing from the insights of Adam Grissom's essay on military innovation. See Adam Grissom, "The Future of Military Innovation Studies," *Journal of Strategic Studies* 29, no. 5 (2006): 905–34.
8. Daniele Rotolo, Diana Hicks, and Ben R. Martin, "What Is an Emerging Technology?" *Research Policy* 44, no. 10 (2015): 1827–43.
9. Frank L. Smith III, "Quantum Technology Hype and National Security," *Security Dialogue* 51, no. 5 (October 2020): 499–516.
10. Karl E. Weick, Kathleen M. Sutcliffe, and David Obstfeld, "Organizing and the Process of Sensemaking," *Organization Science* 16, no. 4 (2005).
11. Quoted by Karl E. Weick, *Sensemaking in Organizations*, vol. 3 (New York: Sage, 1995).
12. Ellie Bartels is among the few wargaming scholars to address sensemaking in her analysis of the varying philosophies of science for research wargames. See Elizabeth

Bartels, "The Science of Wargames: A Discussion of Philosophies of Science for Research Games," paper presented at workshop on War Gaming and Implications for International Relations Research, July 2019.

13. Sally Maitlis and Marlys Christianson, "Sensemaking in Organizations: Taking Stock and Moving Forward," *Academy of Management Annals* 8, no. 1 (2014).

14. Roy Rosenzweig, "Wizards, Bureaucrats, Warriors, and Hackers: Writing the History of the Internet," *American Historical Review* 103, no. 5 (1998): 1530–52.

15. Brandon Valeriano and Ryan C. Maness, *Cyber War versus Cyber Realities: Cyber Conflict in the International System* (New York: Oxford University Press, 2015).

16. August Cole and P. W. Singer, "Thinking the Unthinkable with Useful Fiction," *Journal of Future Conflict*, Fall 2020, 1–13.

17. Jacquelyn Schneider, Benjamin Schechter, and Rachael Shaffer, "A Lot of Cyber Fizzle but Not a Lot of Bang: Evidence about the Use of Cyber Operations from Wargames," *Journal of Global Security Studies* 7, no. 2 (2022): 5.

18. John Curry, "Cyber Wargaming: Practical Ideas from Practical People," lecture presented at Connections Wargaming Conference 2021, June 22, 2021.

19. Nina Kollars and Amanda Rosen, "Bootstrapping and Portability in Simulation Design," *International Studies Perspectives* 17, no. 2 (2016): 202–13.

20. Phillip E. Pournelle, "The Use of M&S and Wargaming to Address Wicked Problems," in *Simulation and Wargaming*, edited by Charles Turnista, Curtis Blais, and Andreas Tolk (New York: Wiley, 2021), 203–21.

21. Benjamin Schechter, "Wargaming Cyber Security," *War on the Rocks*, September 4, 2020, https://warontherocks.com/2020/09/wargaming-cyber-security/.

22. Particularly powerful here is Patrick Lin's warning about Department of Defense games that draw a distinction between legal, ethical, and policy issues with emerging technology. See Patrick Lin, "Pain Rays and Robot Swarms: The Radical New War Games the DoD Plays," *Atlantic*, April 15, 2013.

INDEX

Note: Information in figures and tables is indicated by page numbers in *italics*.

abstraction, 4, 8–9, 24, 37, 46, 52, 80, 129, 178, 208

adjudication, 11–12, 56–57, 59, 73, 75, 84–85, 87–91, *90*, 132–33

adversary: behavior, 26–27; coercion and, 53, *61*; revealing clandestine capabilities to, 33–34; scripted, 44

Afghanistan, 69

AI. *See* artificial intelligence (AI)

Allen, Thomas R., 110

analysis paralysis, 54

analytical utility, *22*; adversary behavior and, 26–27; contextual realism and, 23; in *Cyber Escalation Game*, 28; in *International Crisis War Game*, 26, *28*; in *Island Intercept*, 27; modeling and, 30, *33*; sampling and, 31; simulation and, *33*; survey experiments and, *33*

anonymity, 193–94, 199–201

Army War College, 177–78, 185

artificial intelligence (AI), 15n17, 41, 164, 206–7

asymmetric capabilities, 58

Atlantic Council, 124–25

attribution, 31, 56, 59, 131, 193–95, *196*, 197–200

Australia, 127, 131–32

availability heuristic, 95–96, 100

backcasting, 83–84

Banks, David, 5

Battleship (commercial game), 181

beta testing, 25. *See also* playtesting

bias: for action, 157; for change, 209; cognitive, 95–96, 100; confirmation, 95–96, 99, 101–2; decreasing, 41; empirical data and, 29; modeling and, 30

Bisson, Christian, 113

BlackEnergy, 99, 140

block-style games, 66–67, 70

board game, 3, 115, 118, 165, 181–84, 188

Buchanan, Ben, 181

Bunch of Guys Sitting Around a Table (BOGSAT), 5

business resilience, 152–53

business wargame, 9, 141, 152–55

buy-in, 89, 129, 158–59

Caffrey, Matthew, 2, 38

Caillois, Roger, 110–12, 118

card game, 25, 27, 55–56, 85–87, *86*, 93n9, 119

Center for Naval Analyses (CNA), 79–92

change, cyber wargames as vehicles for, 159–61

ChatGPT, 206–7

CONTRIBUTORS

DAVID E. BANKS is an international relations scholar who focuses on diplomatic history and practice, and wargaming and conflict simulation. In the fall of 2020 he joined Kings College London as wargaming lecturer and academic director of its Wargaming Network. His wargaming research has investigated the potential use of cyber weapons in future conflicts and counterinsurgency techniques against Boko Haram. Most recently, he designed and executed eight interconnected games on subthreshold Multi-Domain Operations conflict for NATO. He has also designed a number of conflict simulations for use in the classroom. His current wargaming research is focused on determining epistemological standards for evaluating wargames as a research method. He also studies diplomatic practice in international society, with a special emphasis on symbolic and rhetorical diplomacy and on great power politics. His writing has been published in the *International Studies Quarterly*, *Security Studies*, and in media outlets including the *Washington Post*, *Time*, the *Independent*, the *Chicago Tribune*, the *Navy Times*, and *US World News & Report*; and he has appeared on BBC News.

RUBY E. BOOTH is a principal member of the Technical Staff at Sandia National Laboratories, where she serves as a national security systems analyst and cybersecurity subject matter expert. She specializes in the interaction between human behavior and cybersecurity. She received her undergraduate degree from Rhodes College and an MS and PhD from the University of Memphis. She is a nonresident fellow of the Berkeley Risk and Security Lab and the Center for Long-Term Cybersecurity at the University of California, Berkeley. Before starting her work in the national security field, she spent over a decade working and writing in the hobby games industry.

LUIS R. CARVAJAL-KIM is a director with Optiv Security's cyber strategy and transformation practice, where he manages practical risk reduction offerings across the spectrum of cybersecurity domains. He also sits on the Board of Directors of City Year Seattle. Before joining Optiv, he helped lead Deloitte's cyber wargaming, risk quantification, and incident response practices in the United States and Canada, and he also spent over a decade in the US national security and intelligence communities. He received his undergraduate and graduate degrees from Carnegie Mellon University, is an alumnus of the Naval War College and the National Defense University, and participated in the Presidential Management Fellows Program.

CHRIS C. DEMCHAK is Grace Hopper Chair of Cyber Security and a senior cyber scholar at the Cyber & Innovation Policy Institute of the US Naval War College. In her publications and current research on cyberspace as a global, insecure, complex, and conflict-prone "substrate," she takes a sociotechnical–economic systems approach to comparative institutional evolution with emerging technologies; adversaries' cyber / artificial intelligence / machine-learning campaigns; virtual wargaming for strategic/organizational learning; and regional/national/enterprise-wide resilience against complex systems surprise. Her manuscripts in progress are "Cyber Westphalia: States, Great Systems Conflict, and Collective Resilience" and "Cyber Commands: Organizing for Cybered Great Systems Conflict." She has degrees in engineering, economics, and comparative complex organization systems / political science.

JAMES DEMUTH is a military operations analyst at Systems Planning and Analysis, Inc. He previously worked as a wargame designer at the National Defense University, where he focused on creating strategic-level educational wargames and exercises for the United States and partner nation government and military audiences. He received an MA in security studies from Georgetown University and a BA in international affairs from George Washington University.

CHRIS DOUGHERTY is a senior fellow in the Defense Program at the Center for a New American Security (CNAS) and co-lead of the CNAS Gaming Lab. His research areas include defense strategy, strategic assessments, force planning, and wargaming. Before joining CNAS, he served as senior adviser to the US deputy assistant secretary of defense for strategy and force development. During this time, he led a handful of major initiatives, including the development and writing of major sections of the "2018 National Defense Strategy."

MATTHEW DUNCAN is the director of the Electricity Information Sharing and Analysis Center (E-ISAC) Intelligence Group, whose mission is to inform the electricity industry's security decision-makers about threats facing the sector; he also developed and managed the *GridEx* V and VI Executive Tabletops.

Before joining the E-ISAC, he was a program manager at the US Department of Energy, managing sector-specific agency and state government outreach. His previous roles include deputy US political officer for the Helmand Provincial Reconstruction Team in Afghanistan; strategic policy analyst at the Pentagon; and research assistant for a US attorney. He has received GIAC certificates in global critical infrastructure protection and cyber threat intelligence, a master's degree in international relations from the Maxwell School of Citizenship and Public Affairs at Syracuse University, and a bachelor's degree from Saint Joseph's University in Philadelphia; he was also a Fulbright grantee to Germany.

SAFA SHAHWAN EDWARDS is the deputy director of the Atlantic Council's Cyber Statecraft Initiative within the Scowcroft Center for Strategy and Security. In this role, she manages the administration and external communications of the initiative, as well as the *Cyber 9/12 Strategy Challenge*, the initiative's global cyber policy and strategy competition. She received an MA in international affairs with a concentration in conflict resolution from George Washington University's Elliott School of International Affairs and a BA in political science from Miami University of Ohio. She is of Bolivian and Jordanian heritage and speaks Spanish and Arabic.

BETHANY GOLDBLUM an associate professor in the Department of Nuclear Engineering at the University of California, Berkeley, and faculty scientist in the Nuclear Science Division of Lawrence Berkeley National Laboratory. As executive director of the Nuclear Science and Security Consortium, she provides strategic direction for a large team of universities and national laboratories working together to train the next generation of nuclear security experts. She is director of the Bay Area Neutron Group and the Public Policy and Nuclear Threats Boot Camp, and she is founder and director of the Nuclear Policy Working Group. She received a PhD in nuclear engineering from the University of California, Berkeley.

ANDREAS HAGGMAN is head of cyber advocacy at the UK Department for Science, Innovation, and Technology. He received a PhD from Royal Holloway, University of London; his thesis was on using cyber wargames for education. He has been a guest lecturer at the UK Defence Academy, German Command and Staff College, Swedish Defence University, and NATO Centre of Excellence Defense Against Terrorism. He has run wargames with public and private institutions in Europe, the United States, and Australia. Before joining the UK civil service, he gained professional experience in various parts of the private sector, including video games, defense, and insurance.

BENJAMIN M. JENSEN holds a dual appointment as a professor at the Marine Corps University's School of Advanced Warfighting and as a senior fellow for future war, gaming, and strategy at the Center for Strategic and International Studies. He is the author or coauthor of five books, including, most recently, *Information in War: Military Innovation, Battle Networks,* and the *Future of Artificial Intelligence* (Georgetown University Press, 2022); *Cyber Strategy: The Evolving Character of Power and Coercion* (Oxford University Press, 2018); and *Military Strategy in the 21st Century: People, Connectivity, and Competition* (Cambria, 2018). He is a series editor for the disruptive technology and strategy series at Oxford University Press. He has extensive wargaming experience working with the US Congress, US Defense Department, NATO, the Economic Community of West African States, and multiple *Fortune* 100 companies. He recently served as the senior research director for the US Cyberspace Solarium Commission, where he ran its competitive strategy game and was the lead author for its final report. Outside academia, he is a reserve officer in the US Army.

DAVID H. "DJ" JOHNSON has had an unconventional career path, flowing between technology and art and all points in between. He has spent the majority of his career designing and prototyping interactive media systems for a variety of brand leaders and private ventures. In his current role as associate professor of game design and development at the New England Institute of Technology, he is best described as an expert on transformational games. His skills as a digital storyteller, animator, writer, designer, and creative facilitator continue to expand in his practice and research in the philosophy of transformational games.

STEVEN KARPPI is a senior research scientist with more than thirty years' experience at the Center for Naval Analyses. He is an expert on organizational analyses, operations research, and real-world operations analysis, and he has received several awards for his analytical excellence and meritorious civilian service in direct support of the US Fleet. He has a broad analytical background, including studies in submarine and antisubmarine warfare; intelligence, surveillance, and reconnaissance programs; Navy integrated fires; platform operational test and evaluation; intelligence community technical analysis; and cyberspace operations analysis. He received a PhD in theoretical physics from George Washington University. He also received BS and MS degrees in electrical engineering from the University of Virginia.

NINA A. KOLLARS is an associate professor at the Cyber & Innovation Policy Institute (CIPI) within the US Naval War College. She served as a special adviser to Heidi Shyu, then the under secretary of defense for research and engineering. Kollars is a scholar of future warfighting, military technological change, innovation, cybersecurity, and cyber warfare / information operations. She also provides analysis on cyber wargaming and education. She received a PhD in political science from The Ohio State University and an MA in international affairs from the Elliott School at George Washington University. She was a senior analyst for the Congressional Cyber Solarium Commission, and she oversees the CIPI's Gravely Directed Research Program.

KIRAN LAKKARAJU is a principal member of the Technical Staff in the Systems Research and Analysis group at Sandia National Laboratories in California. His research interests revolve around interdisciplinary efforts that bring together the social and computational sciences. His background is in experimental wargaming, artificial intelligence, and computational social science. He received a PhD in computer science from the University of Illinois at Urbana-Champaign.

CATHERINE LEA is a senior research scientist, project director, and game designer on the Center for Naval Analyses' (CNA) Wargaming Team. She directs gaming efforts on emerging capabilities including hypersonics, cyber, and space operations. She is an expert on Japanese security policy and US base politics in Japan. Her fieldwork at CNA includes assignments at Amphibious Group Two, US Fleet Forces, and three years in Yokohama providing analytical support to US Navy commands. She received an MA in national security studies from Georgetown University and a BA in political science and economics from the University of California, Berkeley.

HYONG LEE is a senior policy analyst with the Center for Applied Strategic Learning at the National Defense University, which he joined in 2002. He started government service in 1992 as a Presidential Management Intern (now Presidential Management Fellows). In 1996, he started working for the US Pacific Command, eventually becoming the chief of the Decision Support Branch. He oversaw the development and execution of games and exercises, including bilateral/multinational games with regional partners, joint experimentation, and in support of contingency plans. His recent projects include cyber security, information and influence warfare, and integrating technology into exercises. His published papers include "Educating Future Cyber Strategists Through Wargaming: Options, Challenges, and Gaps," for the Thirteenth International Conference on Cyber Warfare and Security (ICCWS), and the coauthored (with James DeMuth) "Representing Strategic Cyber and Information Warfare in Matrix Games," for the Fifteenth ICCWS.

BENJAMIN C. LEITZEL oversaw senior-leader cyberspace education at the US Army War College, where he led the college's cyber working group. He retired as a colonel after a twenty-eight-year Air Force career that included duty as a bomber electronic warfare officer and weapons systems officer (B-1 and B-52), during which he flew sixteen combat missions over Serbia and Kosovo during Operation Allied Force. He was a plans and policy staff officer at the headquarters of the US European Command and a NATO International Security Assistance Force staff officer in Afghanistan. He has over fifteen years' experience in professional military education at the US Army War College, the Joint Forces Staff College, and the Saudi War College.

ED MCGRADY is an adjunct senior fellow in the Defense Program at the Center for a New American Security, and a key member of the Gaming Lab. He writes, speaks, and teaches on the design of professional games. He also runs a business devoted to using games and game techniques to bring innovative experiences in new areas. In the past, he built and directed a team of ten to twenty analysts at the Center for Naval Analyses devoted to the design and execution of professional games. He has written, taught, and presented on the topic of games and their use in organizational and individual learning. He has designed and run games for many different clients, ranging from the White House to the Department of Agriculture.

JUSTIN PEACHEY is a research scientist on the Gaming and Integration Team at the Center for Naval Analyses (CNA). He has designed and facilitated wargames for both the Joint Staff and the Navy that address a variety of topics, from logistics to information warfare. Before joining CNA, he was a visiting professor of mathematics at Davidson College in North Carolina. He received a PhD and an MS in mathematical sciences from Clemson University and a BS in mathematics and secondary education from Grove City College.

ANDREW REDDIE is the founder and faculty director for the Berkeley Risk and Security Lab and is an associate research professor at the Goldman School of Public Policy of the University of California, Berkeley, where he works on projects related to cybersecurity, nuclear weapons policy, wargaming, and emerging military technologies. He is a Bridging the Gap fellow, a nonresident fellow at the Brute Krulak Center at the Marine Corps University, and a term member of the Council on Foreign Relations. His writing has appeared in *Science*, the *Journal of Peace Research*, the *Journal of Cyber Policy*, and the *Bulletin of the Atomic Scientists*, among other outlets, and he has been supported by the Founder's Pledge Fund, Carnegie Corporation of New York, MacArthur Foundation, and the US Department of Energy's Nuclear Science and Security Consortium.

JASON C. REINHARDT is a national security systems analyst and a distinguished member of the Technical Staff at Sandia National Laboratories. His current work is focused on the development of risk analytic methods for the National Risk Management Center in the Department of Homeland Security's Cybersecurity and Infrastructure Security Agency. His research interests include probabilistic analysis methods; quantitative and nonquantitative approaches to risk analysis and management; and modeling and analysis of strategic interaction in conflict escalation, asymmetric deterrence, and stability. He received a PhD in risk analysis from the Department of Management Science and Engineering in Stanford University's School of Engineering. He also received an MS in electrical engineering from Stanford and a BS in electrical engineering from Purdue University's School of Electrical Engineering.

RACHAEL SHAFFER is a researcher in the US Naval War College's Strategic and Operational Research Department. She led logistics and project management for two of the Naval War College's cyber and critical infrastructure games. She was a publication editor for the Naval War College Press's publication, *Ten Years In: Implementing Strategic Approaches to Cyberspace*, as well as for a special issue of *Security Studies* on naval strategy, "Security Studies in a New Era of Maritime Competition." She was also the lead coordinator of the *International Crisis War Game* series at the Hoover Institution at Stanford University.

BENJAMIN H. SCHECHTER is a senior wargaming lead and military operations analyst at Systems Planning and Analysis, Inc., a government contractor. In this role, he leads wargames and projects in support of the Office of the Undersecretary of Defense for Research and Engineering. He is also a nonresident fellow at the US Naval War College. Previously, he was research faculty in the Strategic and Operational Research Department and the cyber wargaming lead for the Cyber & Innovation Policy Institute at the US Naval War College. His research interests include cyber, wargaming, emerging technology, and political psychology.

PAUL SCHMITT is an associate research professor in the Strategy & Operational Research Department of the US Naval War College (NWC). He has designed, developed, executed, and participated in the Defense Department's wargames for over thirty years, and he uses wargaming as the primary methodology for joint professional military education of midcareer officers within the Halsey Alfa Advanced Research Project at the NWC. He is a core member of the Cyber & Innovation Policy Institute. He is a retired submarine officer with thirty years of uniformed service whose last assignment was as deputy director for plans, policy, and resources for the Navy in Europe and Africa. Before joining the NWC's faculty, he was the senior experimentation manager for a broad range of innovation initiatives at the Navy Warfare Development Command. He received an MA in national security and strategic studies from the NWC, and a BA in biological science / marine ecology from Cornell University.

JACQUELYN SCHNEIDER is a Hoover Fellow at the Hoover Institution at Stanford University, is an affiliate of Stanford's Center for International Security and Arms Control, and was previously an assistant professor at the Naval War College's Cyber & Innovation Policy Institute. She has served as a senior policy adviser to the Cyberspace Solarium Commission and has almost two decades of military service as a US Air Force active duty and reserve officer. Her research focuses on the intersection of technology, national security, and political psychology, with a special interest in cybersecurity, unmanned technologies, and military innovation. She received a BA from Columbia University, an MA from Arizona State University, and a PhD from George Washington University.

JEREMY SEPINSKY is the lead wargame designer at the Center for Naval Analyses. He has designed and facilitated dozens of wargames at the Navy and Joint commands, as well as for the Office of the Under Secretary of Defense–Policy and the Joint Staff. His recent wargames have covered topics such as logistics, personnel organization, cyberspace operations, national strategy, technology planning, special operations, and homeland defense. Before joining the Center for Naval Analyses, he was an assistant professor of physics and astronomy at the University of Scranton. He received a PhD and an MS in physics and astronomy from Northwestern University, and a BS in astronomy and astrophysics from Villanova University.

FRANK L. SMITH III is a professor and director of the Cyber & Innovation Policy Institute at the US Naval War College. His interdisciplinary research and teaching examine how ideas about emerging technology—especially bad ideas—influence national security and international relations. His previous work includes his book *American Biodefense*, as well as articles published in *Security Studies*, *Social Studies of Science*, *Security Dialogue*, *Health Security*, *Asian Security*, and *The Lancet*. He received a PhD in political science and a BS in biological chemistry from the University of Chicago.

Printed in the USA
CPSIA information can be obtained
at www.ICGtesting.com
LVHW041328201123
R17973100001B/R179731PG763548LVX00001B/1